Molecular Methods of Plant Analysis

Editors:
J.F. Jackson (Managing Editor)
H.F. Linskens
R.B. Inman

Volume 22

Springer

Berlin
Heidelberg
New York
Barcelona
Hong Kong
London
Milan
Paris
Tokyo

Volumes Already Published in this Series
(formerly "Modern Methods of Plant Analysis"):

Volume 1:   Cell Components
            1985, ISBN 3-540-15822-7
Volume 2:   Nuclear Magnetic Resonance
            1986, ISBN 3-540-15910-X
Volume 3:   Gas Chromatography/Mass Spectrometry
            1986, ISBN 3-540-15911-8
Volume 4:   Immunology in Plant Sciences
            1986, ISBN 3-540-16842-7
Volume 5:   High Performance Liquid Chromatography in Plant Sciences
            1986, ISBN 3-540-17243-2
Volume 6:   Wine Analysis
            1988, ISBN 3-540-18819-3
Volume 7:   Beer Analysis
            1988, ISBN 3-540-18308-6
Volume 8:   Analysis of Nonalcoholic Beverages
            1988, ISBN 3-540-18820-7
Volume 9:   Gases in Plant and Microbial Cells
            1989, ISBN 3-540-18821-5
Volume 10: Plant Fibers
            1989, ISBN 3-540-18822-3
Volume 11: Physical Methods in Plant Sciences
            1990, ISBN 3-540-50332-3
Volume 12: Essential Oils and Waxes
            1991, ISBN 3-540-51915-7
Volume 13: Plant Toxin Analysis
            1992, ISBN 3-540-52328-6
Volume 14: Seed Analysis
            1992, ISBN 3-540-52737-0
Volume 15: Alkaloids
            1994, ISBN 3-540-52738-9
Volume 16: Vegetables and Vegetable Products
            1994, ISBN 3-540-55843-8
Volume 17: Plant Cell Wall Analysis
            1996, ISBN 3-540-59406-X
Volume 18: Fruit Analysis
            1995, ISBN 3-540-59118-4
Volume 19: Plant Volatile Analysis
            1997, ISBN 3-540-61589-X
Volume 20: Analysis of Plant Waste Materials
            1999, ISBN 3-540-64669-8
Volume 21: Analysis of Taste and Aroma
            2002, ISBN 3-540-41753-2
Volume 22: Testing for Genetic Manipulation in Plants
            2002, ISBN 3-540-43153-5

Testing for Genetic Manipulation in Plants

Edited by
J.F. Jackson and H.F. Linskens

With 33 Figures, 2 in Color
and 16 Tables

Springer

Professor J.F. Jackson
Department of Horticulture
Viticulture and Oenology
University of Adelaide
Waite Campus
SA 5064 Glen Osmond
Australia

Professor H.F. Linskens
Goldberglein 7
91056 Erlangen
Germany

Professor R.B. Inman
Institute of Molecular Virology
University of Wisconsin-Madison
Robert M. Bock Laboratories
1525 Linden Drive
Madison, Wisconsin 53706-1596
USA

ISSN 1619-5221
ISBN 3-540-43153-5 Springer-Verlag Berlin Heidelberg New York

Library of Congress Cataloging-in-Publication Data

Testing for genetic manipulation in plants / edited by J.F. Jackson, H.F. Linskens, and
R.B. Inman.
    p.   cm. – (Molecular methods of plant analysis ; v. 22)
    Includes bibliographical references (p.).
    ISBN 3-540-43153-5 (alk. paper)
    1. Plant genetic transformation.   2. Plant genetic engineering.   3. Crops – Genetic
engineering.   I. Jackson, J.F. (John F.), 1935–   II. Linskens, H.F. (Hans F.) 1921–
III. Inman, R.B. (Ross B.)   IV. Series.

QK865 .M57 vol. 22
[SB123.57]
571.2'028 s – dc21
[581.3'5]   2002022544

Springer-Verlag Berlin Heidelberg New York
a member of BertelsmannSpringer Science+Business Media GmbH
http://www.springer.de

© Springer-Verlag Berlin Heidelberg 2002
Printed in Germany

The use of general descriptive names, registered names, trademarks, etc. in this publication does not
imply, even in the absence of a specific statement, that such names are exempt from the relevant pro-
tective laws and regulations and therefore free for general use.

Cover design: Design & Production GmbH, Heidelberg
Typesetting: SNP Best-set Typesetter Ltd., Hong Kong
SPIN 10854590   31/3130/Wi – 5 4 3 2 1 0 – Printed on acid-free paper

Preface

Molecular Methods of Plant Analysis

Concept of the Series

The powerful recombinant DNA technology and related developments have had an enormous impact on molecular biology. Any treatment of plant analysis must make use of these new methods. Developments have been so fast and the methods so powerful that the editors of *Modern Methods of Plant Analysis* have now decided to rename the series *Molecular Methods of Plant Analysis*. This will not change the general aims of the series, but best describes the thrust and content of the series as we go forward into the new millennium. This does not mean that all chapters a priori deal only with the methods of molecular biology, but rather that these methods are to be found in many chapters together with the more traditional methods of analysis which have seen recent advances. The numbering of the volumes of the series therefore continues on from 20, which is the most recently published volume under the title *Modern Methods of Plant Analysis*.

As indicated for previous volumes, the methods to be found in *Molecular Methods of Plant Analysis* are described critically, with hints as to their limitations, references to original papers and authors being given, and the chapters written so that there is little need to consult other texts to carry out the methods of analysis described. All authors have been chosen because of their special experience in handling plant material and/or their expertise with the methods described. The volumes of the series published up to now fall into three groups: Volumes 1–5 and Volume 11 dealing with some basic principles of methods, Volumes 6, 7, 8, 10, 14, 16, 18 and 20 being a group determined by the raw plant material being analysed, and a third group comprising Volumes 9, 12, 13, 15, 17 and 19 which are separated from the other volumes in that the class of substances being analysed for is indicated in the volume title. Volume 21 and future volumes of *Molecular Methods of Plant Analysis* will continue in a similar vein but will include more chapters involved with the methods of molecular biology.

Development of the Series

The handbook, *Modern Methods of Plant Analysis*, was first introduced in 1954, and was immediately successful, seven volumes appearing between 1956 and 1964. This first series was initiated by Michael Tracey of Rothamsted and Karl Paech of Tübingen. The so-called *New Series of Modern Methods of Plant Analysis*, Volumes 1–20, began in 1985 and has been edited by Paech's successor, H.F. Linskens of Nijmegen, The Netherlands, and John F. Jackson of Adelaide, South Australia. These same editors have now teamed up with a third, Ross B. Inman of Madison, Wisconsin, USA, to produce the renamed series *Molecular Methods of Plant Analysis*. As before, the editors are convinced that there is a real need for a collection of reliable, up-to-date methods of plant analysis covering large areas of applied biology ranging from agricultural and horticultural enterprises to pharmaceutical and technical organizations concerned with material of plant origin.

Future volumes will include such topics as Genetic Transformation of Plants and Various Aspects of Plant Genomics.

Volume 22: *Testing for Genetic Manipulation in Plants*

This second volume in the "molecular" series takes up the topic of testing for genetic manipulation (GM) by the newly developed recombinant DNA technology in plants. This volume includes methods for reporter genes which enable the researcher to determine whether or not a procedure has introduced the desired gene into a transgenic plant, and can also provide means for others, if these genes have not been removed, to detect GM in plants or plant products suspected of being genetically manipulated. Chapters are included which deal with the removal of these and other marker genes from GM plants, as well as chapters devoted to safety assessment of various plants and their products, and to procedures characterizing chromosomal aberrations which can arise in transgenic plants.

The first chapter expounds on the topic of selectable and screenable markers used in GM rice and sets out methods for these. Since the safety of marker or reporter genes involving resistance to antibiotics or herbicides has aroused public concern in many countries, it is important to be able to test for their presence. The availability of tests for these genes becomes even more important when it is considered that antibiotic-resistance markers could be a health threat if they escaped into crops in general use, while herbicide-resistance markers could pose a real problem to the ecosystem if they escaped into the weed population. Following this first chapter, the next three chapters deal with methods for the three most commonly used reporter genes: the green fluorescent protein (GFP) gene, the luciferase (LUC) gene and the β-glucuronidase (GUS) gene. Each of these has its own advantages, often to do with the simplicity of detection, desirability of destruction of the plant or the need for quantification of genetically manipulated cells.

Methods for detection of "GM" in grain legumes are to be found in the next chapter, concentrating on the use of sensitive qualitative and quantitative PCR-based techniques. While ELIZA assays have been used in the past for detecting the various gene products in genetically manipulated plants or plant products and they are cheaper; nevertheless, the PCR-based techniques are much more sensitive and increasingly used, for example, in the testing of seeds suspected of containing some GM material. Strategies have been developed to eliminate markers from transgenic plants. This topic is the subject of the next two chapters, since it is thought that, in the future, it would be prudent to alleviate perceived risks by eliminating selectable marker genes from transgenic crops prior to their field release and commercialization. In any testing for GM, it is essential that one knows if any of the markers have been removed, and what other genetic material associated with the introduction of the transgene remains or has been removed. The first of these chapters deals with cotransformation, transposon-mediated approaches, site-specific recombination and intrachromosomal homologous recombination as strategies to generate marker-free transgenic plants, while the second chapter concentrates on a GST-MAT vector method.

Safety Aassessment methods are essential in dealing with GM plants or plant products that are to be consumed by animals or humans. Thus two chapters are included on this topic, the first dealing with the safety assessment of insect-protected corn or cotton crops to be used for feeding poultry, swine or cattle, while the second includes safety assessment methods for genetically modified rice and potatoes. Finally, the volume is rounded off with two chapters dealing with methods designed to examine chromosomal and genetic aberrations in GM soybean, in one case, and GM barley, in another. It is not always appreciated that transformation technologies can lead to gross cytological changes in chromosomes.

J.F. JACKSON, Managing Editor, H.F. LINSKENS, R.B. INMAN

Contents

1 Selectable and Screenable Markers for Rice Transformation
R.M. TWYMAN, E. STÖGER, A. KOHLI, T. CAPELL, and P. CHRISTOU

1.1 Introduction 1
1.2 Dominant Selectable Markers for Rice 3
    1.2.1 Aminoglycoside 3′-Phosphotransferase
        (Neomycin Phosphotransferase) 3
    1.2.2 Hygromycin Phosphotransferase 4
    1.2.3 Phosphinothricin Acetyltransferase 4
    1.2.4 Other Dominant Selectable Markers 5
1.3 Novel Selectable Markers 7
    1.3.1 Innocuous Markers 7
    1.3.2 Counterselectable Markers 7
1.4 Screenable Marker Genes 8
    1.4.1 β-Glucuronidase (*gusA*) 9
    1.4.2 Firefly Luciferase (*luc*) 11
    1.4.3 Green Fluorescent Protein (*gfp*) 11
1.5 Strategies for Marker-Gene Delivery 13
References 14

2 Use of Green Fluorescent Protein to Detect Transformed Shoots
J. MOLINIER and G. HAHNE

2.1 Introduction 19
2.2 GFP: Suitable as a Visually Selectable Marker *In Planta*? 20
    2.2.1 Important Properties of the Protein 20
    2.2.2 Properties of a Useful Selectable Marker in Plant
        Transformation Technology 21
2.3 GFP Expression and Detection in Primary
    Transformed Tissues 22
    2.3.1 Transient Expression and GFP Detection 22
    2.3.2 Detection Equipment and Troubleshooting 22
    2.3.3 Stable Expression and GFP Detection in Primary
        Transformed Tissues 23
2.4 GFP for Screening of Segregating Populations 28
2.5 Conclusion 29
References 29

3 Luciferase Gene Expressed in Plants, Not in *Agrobacterium*
S.L. MANKIN

3.1   Introduction 31
3.2   Preventing Bacterial Expression 31
3.3   Imaging Luciferase Activity In Planta 32
3.4   Measuring Luciferase Activity in Plant Extracts 34
References 35

**4 Use of *β*-Glucuronidase (GUS) To Show Dehydration
and High-Salt Gene Expression**
K. NAKASHIMA and K. YAMAGUCHI-SHINOZAKI

4.1   Introduction 37
4.2   What Is GUS? 39
4.3   Trasgenic Plants Carrying Promoter-*GUS* Constructs 41
       4.3.1   Construction of Promoter-*GUS* Fusion Genes 44
       4.3.2   Introduction of Promoter-*GUS* Constructs into
               Agrobacterium 45
       4.3.3   Transformation of Plants with *Agrobacterium* 46
               4.3.3.1 Transformation of *Arabidopsis* Plants 46
4.4   Fluorometric Assay 49
       4.4.1   Introduction 49
       4.4.2   Stress Conditions 49
               4.4.2.1 Plant Preparation 49
               4.4.2.2 Dehydration 49
               4.4.2.3 High Salinity 49
               4.4.2.4 ABA Treatment 50
               4.4.2.5 Other Treatments 50
       4.4.3   Protein Assay 50
       4.4.4   Sample Preparation 50
       4.4.5   Fluorometric Assay 51
4.5   Histochemistry 51
       4.5.1   Introduction 51
       4.5.2   Histochemistry 52
4.6   Northern Analysis of GUS 52
       4.6.1   Introduction 52
       4.6.2   RNA Extraction 53
       4.6.3   RNA Blotting 54
       4.6.4   Northern Hybridization 55
4.7   Application of the GUS System 56
       4.7.1   Transient Assay 56
       4.7.2   Transactivation Experiment 57
       4.7.3   Promoter Tagging (Enhancer Trap) 57
4.8   Conclusion 57
References 59

5 Methods for Detecting Genetic Manipulation in Grain Legumes
H.-J. JACOBSEN and R. GREINER

5.1   Introduction   63
5.2   Detection at the DNA Level   64
5.3   PCR Analysis   65
5.4   Control PCR and Specific PCR Systems   66
5.5   Quantitative Approach   69
5.6   Competitive PCR   69
5.7   Real-Time PCR Systems   70
5.8   Concluding Remarks   70
References   71

6 Elimination of Selectable Marker Genes from Transgenic Crops
A.P. GLEAVE

6.1   Introduction   73
6.2   Co-transformation   74
6.3   Transposon-Mediated Approaches   78
      6.3.1   Transposon-Mediated Repositioning   78
      6.3.2   Transposon-Mediated Elimination   80
6.4   Site-Specific Recombination   81
      6.4.1   The Cre/loxP System   82
      6.4.2   The FLP/frt System   86
      6.4.3   The R/RS System   88
6.5   Intrachromosomal Homologous Recombination   89
6.6   Conclusions and Future Prospects   90
References   91

7 GST-MAT Vector for the Efficient and Practical Removal of Marker Genes from Transgenic Plants
H. EBINUMA, K. SUGITA, E. MATSUNAGA, S. ENDO, and K. YAMADA

7.1   Introduction   95
7.2   ipt-Type MAT Vectors   96
      7.2.1   Transposable Element   96
      7.2.2   Site-Specific Recombination System   97
      7.2.3   Advantages of the ipt Gene 101
7.3   Two-Step Transformation 103
      7.3.1   Promoter of the R Gene 103
      7.3.2   Promoter of the ipt Gene 105
      7.3.3   Combination of the ipt and iaaM/H Genes 108
      7.3.4   Transgene Stacking 110
7.4   Single-Step Transformation 112

7.5   Cloning Vector for Desired Genes 114
7.6   Concluding Remarks 115
References 116

8 Safety Assessment of Insect Protected Crops: Testing the Feeding Value of *Bt* Corn and Cotton Varieties in Poultry, Swine and Cattle
B. HAMMOND, E. STANISIEWSKI, R. FUCHS, J. ASTWOOD, and G. HARTNELL

8.1   Introduction 119
      8.1.1   Food Safety Standards 119
      8.1.2   Testing for Food and Feed Safety 120
8.2   Insect Protection Traits 122
8.3   Benefits 123
8.4   Safety Assessment of the Cry Insect-Control Proteins 124
8.5   Mode of Action 125
8.6   Substantial Equivalence Based on Compositional Analysis 126
8.7   Current Products 127
8.8   Grower Acceptance 127
8.9   Future Products 127
8.10 Farm-Animal Studies 128
      8.10.1 *Bt* Corn 128
              8.10.1.1 Poultry 128
              8.10.1.2 Lactating Cows 131
              8.10.1.3 Beef and Sheep 131
              8.10.1.4 Swine 132
8.11 Cottonseed 133
8.12 Conclusions 134
References 135

9 Safety Assessment of Genetically Modified Rice and Potatoes with Soybean Glycinin
K. MOMMA, W. HASHIMOTO, S. UTSUMI, and K. MURATA

9.1   Introduction 139
9.2   Safety Assessment of Genetically Modified Crops 140
      9.2.1   Genetically Modified Rice 141
      9.2.2   Genetically Modified Potatoes 144
9.3   Concluding Remarks 146
References 149

10 Chromosomal and Genetic Aberrations in Transgenic Soybean
R.J. SINGH

10.1 Introduction 153
10.2 Times in Culture with 2,4-D Prior to Transformation 154
10.3 Genetic Background of the Explants 159

10.4 Seed Fertility in Transgenic Soybean 160
10.5 Cytological Basis of Gene Silencing 163
10.6 Conclusions 164
References 165

**11 Transgenic Barley (*Hordeum vulgare* L.)
and Chromosomal Variation**
M.-J. CHO, H.W. CHOI, P. BREGITZER, S. ZHANG, and P.G. LEMAUX

11.1 Introduction 169
11.2 Chromosomal Variation in Nontransgenic Barley Plants 170
11.3 Chromosomal Variation in Transgenic Barley Plants 172
11.4 Fidelity and Quality of Transgenic Barley Plants 177
    11.4.1 Comparative Analysis of Genomic Stability in Plants
           Derived from Tissues Generated Using Different in Vitro
           Proliferation Processes 177
    11.4.2 Somaclonal Variation and Field Performance
           of Transgenic Plants Derived from Embryogenic
           Callus 179
    11.4.3 Stability of Transgenes and Transgene Expression 180
11.5 Conclusions and Future Perspectives 184
References 185

Subject Index 189

List of Contributors

J. Astwood
Monsanto C2SE, 800 North Lindbergh Boulevard, St. Louis, Missouri 63167, USA

P. Bregitzer
USDA-ARS, National Small Grains Germplasm Research Center, P.O. Box 307, Aberdeen, Idaho 83210, USA

T. Capell
Molecular Biotechnology Unit, John Innes Centre, Norwich, UK

M.-J. Cho
Department of Plant and Microbial Biology, University of California, Berkeley, California 94720, USA

H.W. Choi
Department of Plant and Microbial Biology, University of California, Berkeley, California 94720, USA

P. Christou
Molecular Biotechnology Unit, John Innes Centre, Norwich, UK

H. Ebinuma
Pulp and Paper Research Laboratory, Nippon Paper Industries Co., Ltd., 5-21-1, Oji, Kita-ku, Tokyo 114-0002, Japan

S. Endo
Pulp and Paper Research Laboratory, Nippon Paper Industries Co., Ltd., 5-21-1, Oji, Kita-ku, Tokyo 114-0002, Japan

R. Fuchs
Monsanto C2SE, 800 North Lindbergh Boulevard, St. Louis, Missouri 63167, USA

A.P. Gleave
Plant Health and Development Group, The Horticulture and Food Research Institute of New Zealand Ltd., Private Bag 92169, Auckland, New Zealand

R. GREINER
Lehrgebiet Molekulargenetik, Universität Hannover, Herrenhäuser Strasse 2, 30419 Hannover, Germany

G. HAHNE
Institut de Biologie Moléculaire des Plantes, CNRS et ULP, 12 rue du Général Zimmer, 67084 Strasbourg, Cedex, France

B. HAMMOND
Monsanto C2SE, 800 North Lindbergh Boulevard, St. Louis, Missouri 63167, USA

G. HARTNELL
Monsanto C2SE, 800 North Lindbergh Boulevard, St. Louis, Missouri 63167, USA

W. HASHIMOTO
Basic and Applied Molecular Biotechnology, Division of Food and Biological Science, Graduate School of Agriculture, Kyoto University, Uji, Kyoto 611-0011, Japan

H.-J. JACOBSEN
Lehrgebiet Molekulargenetik, Universität Hannover, Herrenhäuser Strasse 2, 30419 Hannover, Germany

A. KOHLI
Molecular Biotechnology Unit, John Innes Centre, Norwich, UK

P.G. LEMAUX
Department of Plant and Microbial Biology, University of California, Berkeley, California 94720, USA

S.L. MANKIN
BASF Plant Science LLC, 26 Davis Drive, Research Triangle Park, North Carolina 27709, USA

E. MATSUNAGA
Pulp and Paper Research Laboratory, Nippon Paper Industries Co., Ltd., 5-21-1, Oji, Kita-ku, Tokyo 114-0002, Japan

J. MOLINIER
Friedrich Mieschner-Institut, Maulbeer Strasse 66, CH-4058 Basel, Switzerland

K. MOMMA
Basic and Applied Molecular Biotechnology, Division of Food and Biological Science, Graduate School of Agriculture, Kyoto University, Uji, Kyoto 611-0011, Japan

K. Murata
Basic and Applied Molecular Biotechnology, Division of Food and Biological Science, Graduate School of Agriculture, Kyoto University, Uji, Kyoto 611-0011, Japan

K. Nakashima
Biological Resources Division, Japan International Research Center for Agricultural Sciences (JIRCAS), 1-2 Ohwashi, Tsukuba, Ibaraki 305-8686, Japan

R.J. Singh
Department of Crop Sciences, University of Illinois, 1102 South Goodwin Avenue, Urbana, Illinois 61801, USA

E. Stanisiewski
Monsanto C2SE, 800 North Lindbergh Boulevard, St Louis, Missouri 63167, USA

E. Stöger
Molecular Biotechnology Unit, John Innes Centre, Norwich, UK

K. Sugita
Pulp and Paper Research Laboratory, Nippon Paper Industries Co., Ltd., 5-21-1, Oji, Kita-ku, Tokyo 114-0002, Japan

R. M. Twyman
Molecular Biotechnology Unit, John Innes Centre, Norwich, UK

S. Utsumi
Basic and Applied Molecular Biotechnology, Division of Food and Biological Science, Graduate School of Agriculture, Kyoto University, Uji, Kyoto 611-0011, Japan

K. Yamada
Pulp and Paper Research Laboratory, Nippon Paper Industries Co., Ltd., 5-21-1, Oji, Kita-ku, Tokyo 114-0002, Japan

K. Yamaguchi-Shinozaki
Biological Resources Division, Japan International Research Center for Agricultural Sciences (JIRCAS), 1-2 Ohwashi, Tsukuba, Ibaraki 305-8686, Japan

S. Zhang
Department of Plant and Microbial Biology, University of California, Berkeley, California 94720, USA

1 Selectable and Screenable Markers for Rice Transformation

R.M. Twyman, E. Stöger, A. Kohli, T. Capell, and P. Christou

1.1 Introduction

Rice transformation is a major goal in cereal biotechnology, not only because rice is the world's most important food crop but also because this species has now been recognized as the model for cereal genomics. A number of alternative transformation strategies are available, the most widely used of which are particle bombardment (Christou et al. 1991) and *Agrobacterium*-mediated transformation (Hiei et al. 1994). Regardless of the chosen strategy, transformation is a low-efficiency process. Since most foreign genes introduced into plants do not confer a phenotype that can be used conveniently for the identification or selective propagation of transformed cells, marker genes must therefore be introduced along with the transgene of interest to provide such a phenotype.

A selectable marker gene encodes a product that allows the transformed cell to survive and grow under conditions that kill or restrict the growth of nontransformed cells. Most such genes used in rice are dominant selectable markers that confer resistance to antibiotics or herbicides (Table 1.1). In recent years, public concern over the safety of such markers has been aroused in some parts of Europe. It is perceived that antibiotic-resistance markers could pose a threat to health if they "escaped" into nature (especially those conferring resistance to antibiotics used in clinical medicine), while herbicide-resistance markers could damage the ecosystem if they spread from crops to related weed species. Although research suggests that the likelihood of such horizontal gene transfer is minimal, some laboratories have focused on the development of a new breed of selectable markers whose biological activities pose no potential harmful effects. Strategies have also been developed to eliminate markers from transgenic plants. These include attempts to introduce the transgene of interest and the marker gene at separate loci, so they can be separated by sexual crossing (Komari et al. 1996), and the use of site-specific recombination systems such as Cre-*loxP* (Dale and Ow 1991; Zubko et al. 2000) or transpositional mechanisms involving *Ac-Ds* elements (Goldsbrough et al. 1993) to eliminate the markers once the transgenic line is stable.

A screenable marker gene, also called a visual marker, scorable marker or reporter gene, generates a product that can be detected using a simple and often quantitative assay (Table 1.2). Such markers are used for a variety of purposes, e.g. confirming transformation, determining transformation efficiency

Molecular Methods of Plant Analysis, Vol. 22
Testing for Genetic Manipulation in Plants
Edited by J.F. Jackson, H.F. Linskens, and R.B. Inman
© Springer-Verlag Berlin Heidelberg 2002

Table 1.1. Selectable markers and selective agents that have been used or evaluated in rice

Selectable marker	Source	Principle of selection	Comments and principle references for use in rice
ble (glycopeptide-binding protein)	*Streptalloteichus hindustantus*	Confers resistance to the glycopeptide antibiotics bleomycin and pheomycin (and the derivative Zeocin)	Bleomycin evaluated for selection against wild-type cells. *ble* gene not yet introduced as marker; Dekeyser et al. (1989)
dhfr (dihydrofolate reductase)	Mouse	Confers resistance to methotrexate	Tested in cell suspension cultures; Meijer et al. (1991)
hpt (hygromycin phosphotransferase)	*Klebsiella* spp.	Confers resistance to hygromycin B	Christou et al. (1991); Hiei et al. (1994)
nptII (neomycin phosphotransferase)	*Escherichia coli*	Confers resistance to the aminoglycoside antibiotics neomycin, kanamycin and G148 (geneticin)	Toriyama et al. (1988); Raineri et al. (1990); Chan et al. (1993)
bar and *pat* (phosphinothricin acetyltransferase)	*Streptomyces hygroscopicus*	Confers resistance to phosphinothricin (glufosinate) and the herbicides bialaphos and Basta	Christou et al. (1991); Cao et al. (1992)
csr1–1 (acetolactate synthase)	*Arabidopsis thaliana*	Confers resistance to chlorsulphuron	Li et al. (1992)
tms2 (indoleacetic acid hydrolase)	*Agrobacterium tumefaciens*	Confers sensitivity to naphthaleneacetamide (NAM)	Uphadyaya et al. (2000)

Table 1.2. Screenable markers used in rice

Screenable marker	Source	Principle of assay	Comments and principle references for use in rice
gusA (β-glucuronidase, GUS)	*Escherichia coli*	Catalyses hydrolysis of β-glucuronides; variety of colorimetric, fluorometric and chemiluminescent assay formats; can be used for in vitro and in vivo nondestructive assays	Sensitive and versatile. Stable protein; Christou et al. (1991)
luc (firefly luciferase)	The firefly *Photinus pyralis*	Light produced in the presence of luciferase, its substrate luciferin, Mg^{2+}, oxygen and ATP	Sensitive; unstable protein useful for inducible/repressible systems; Baruah-Wolff et al. (1999)
gfp (green fluorescent protein)	The jellyfish *Aqueorea victoria*	Spontaneous fluorescence under UV or blue light.	Sensitive, substrate independence means it can be used in living plant cells for real-time imaging; Vain et al. (1998)

and monitoring gene or protein activity. The ability of screenable markers to form fusion genes at the transcriptional level can be used to assay the activity of regulatory elements. This property can be exploited, in the form of gene-trap vectors, to isolate and characterize new genes as part of a functional genomics strategy. Screenable markers can also form translational fusion products, which allows them to be used to monitor protein localization in the cell or at a whole-plant level.

We begin this chapter by considering the benefits and potential disadvantages of individual marker genes used in transgenic rice. It is desirable to have available a range of different markers so that techniques such as multi-step transformation (with different selectable markers), or the simultaneous monitoring of the activities of several genes (with different visual markers) are possible. We conclude with a brief discussion of strategies for marker-gene delivery and expression.

1.2 Dominant Selectable Markers for Rice

1.2.1 Aminoglycoside 3'-Phosphotransferase (Neomycin Phosphotransferase)

The first reports of gene transfer to rice involved the transformation of protoplasts, either by electroporation or mediated by polyethylene glycol (Zhang et al. 1988; Zhang and Wu 1988; Datta et al. 1990). In these studies, the selectable marker gene *nptII* (also known as *aph(3')II, kan* and *neo*) was used, conferring resistance to aminoglycoside antibiotics such as neomycin, kanamycin and geneticin (G418).

Kanamycin is an effective selective agent for transformed rice protoplasts, but rice callus shows natural resistance to this antibiotic and survives on medium containing levels of kanamycin up to ten times higher than that sufficient to kill many other species (Caplan et al. 1992). It has also been found that protoplast-derived callus selected on kanamycin is very inefficient in terms of regeneration, and that a large number of albino plants arise from such experiments (Toriyama et al. 1988). The earliest reports of rice transformation using *Agrobacterium* also involved kanamycin selection. Raineri et al. (1990) used 200 mg l^{-1} kanamycin to select rice callus derived from mature embryos, but no transgenic plants were recovered.

Selection on G418 is more effective than kanamycin, perhaps because G418 is more toxic than kanamycin to rice cells. Furthermore, G418 selection also results in the recovery of a higher proportion of fertile, green, transgenic plants (Ayres and Park 1994). Even so, the overall efficiency of the procedure is low. Chan et al. (1992, 1993) used this antibiotic in combination with the *nptII* gene following *Agrobacterium*-mediated transformation of rice roots and immature embryos. In both cases, it was possible to generate G418-resistant callus, but transgenic plants could be recovered only when using embryos as the target tissue. However, a total of just four transgenic plants was produced following selection on 40 mg G418 l^{-1}.

1.2.2 Hygromycin Phosphotransferase

The gene for hygromycin phosphotransferase (*hpt* or *aphIV*) confers resistance to aminoglycoside antibiotics such as hygromycin B (van den Elzen et al. 1985). This antibiotic is much more effective than kanamycin and G418 for the selection of transformed rice tissues because there is no innate resistance, thus providing strong discrimination between transformed and nontransformed cells (Christou and Ford 1995). Furthermore, this antibiotic does not appear to inhibit regeneration, nor affect the fertility of transgenic plants (Aldemita and Hodges 1996).

Hygromycin in combination with the *hpt* marker gene has therefore been used as a selection system in a large number of transformation experiments, encompassing both particle bombardment and *Agrobacterium*-mediated procedures. Indeed, *hpt* was used for the first experiments involving the stable transformation of rice by particle bombardment. Immature rice embryos were bombarded and plated on regeneration medium supplemented with 50 mg hygromycin B l^{-1} (Christou et al. 1991). Continuous selection resulted in the appearance of transformed embryogenic callus, from which transgenic plants were recovered. Nontransformed callus, and callus transformed with alternative markers did not survive when plated on hygromycin-supplemented medium (Fig. 1.1).

The first report of high-efficiency transformation of rice using *Agrobacterium* also involved hygromycin selection (Hiei et al. 1994). Callus derived from immature embryos, scutella and suspension cells of various Japonica cultivars was transformed and selected on medium containing between 50 and 100 mg hygromycin l^{-1}, resulting in the recovery of large numbers of fertile transgenic plants. Hygromycin at 50 mg l^{-1} was also used for the first successful *Agrobacterium*-mediated transformation of Indica and Javanica rice cultivars (Dong et al. 1996; Rashid et al. 1996). Efficient recovery of transgenic rice plants following *Agrobacterium*-mediated transformation has also been reported using 30 mg hygromycin l^{-1} (Aldemita and Hodges 1996).

1.2.3 Phosphinothricin Acetyltransferase

The *bar* and *pat* genes from *Streptomyces hygroscopicus* encode phosphinothricin acetyltransferase, an enzyme that provides resistance to phosphinothricin (PPT) and derivatives such as bialaphos, which are competitive inhibitors of glutamine synthesis (de Block et al. 1987). PPT, or more correctly its ammonium salt glufosinate, is the active component of the herbicide Basta.

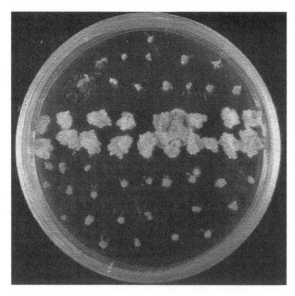

Fig. 1.1. Rice callus selected on 50 mg hygromycin ml^{-1}. As shown, this gives strong discrimination between transformed and non-transformed tissue. (Sudhakar et al.1998)

An important advantage of this selection strategy is that rice plants can be sprayed with the selective agent in the glass-house or in the field. Rice plants transformed with this marker gene can grow when sprayed with up to 2,000 ppm of the herbicide, a concentration that is sufficient to kill nontransformed plants (Fig. 1.2).

The *bar* gene was evaluated as a selectable marker for rice in the earliest particle bombardment experiments and does not adversely affect regeneration (Christou et al. 1991). The effective concentration of PPT ranged from approximately 0.5 to 10 mg l^{-1}. Cao et al. (1992) used *bar* and PPT for the selection of transformed callus generated by particle bombardment, and showed that PPT was at least as effective as hygromycin for the selection of transformed cells. Park et al. (1996) were the first to use *bar* for the selection of rice tissue (in this case shoot apices) transformed by *Agrobacterium*. However, transgenic plants were recovered at a low frequency and the marker was subject to transgene silencing in R2 plants. Basta has also been used in combination with *bar* for the regeneration of transgenic rice plants from transformed protoplasts (Rathore et al. 1993).

1.2.4 Other Dominant Selectable Markers

A wide range of dominant selectable markers is available for plants, including genes providing resistance to other antibiotics (e.g. bleomycin, methotrexate,

Fig. 1.2. Wild-type and transgenic (phosphinothricin-resistant) rice plants after exposure to the herbicide Basta, which contains phosphinothricin. (Oard et al. 1996)

streptomycin and spectinomycin) and herbicides (e.g. glyphosate and chlorsulphuron) (Twyman et al. 2001). Although not widely used in rice, some of these alternative markers have been evaluated (Dekeyser et al. 1989; Caplan et al. 1992), and in the future they may be more widely adopted as the multistep transformation of plants already carrying the commonly used selectable markers becomes more commonplace.

A mutant acetolactate synthase gene, *csr1–1*, isolated from sulfonylurea-resistant *Arabidopsis thaliana*, was introduced into rice protoplasts by Li et al. (1992). Callus derived from the transformed protoplasts was resistant to 200 times the concentration of chlorsulphuron normally lethal to wild-type rice tissue. Therefore, effective selection was possible at levels of the herbicide as low as 10 nM. Fertile transgenic rice plants were recovered from the transformed callus, showing that the acetolactate synthase gene was compatible with efficient regeneration.

The murine dihydrofolate reductase gene has also been evaluated as a selectable marker, since it has been used effectively for the recovery of other transgenic crop plants, including maize (Golovkin et al. 1993). The gene was introduced into rice protoplasts, and integration into the genomic DNA was confirmed by Southern blot analysis, the presence of *dhfr* RNA, and resistance of the cells to high levels of the drug methotrexate added to the culture medium (Meijer et al. 1991). Methotrexate has been compared to the more commonly used selective agents such as hygromycin, PPT, and G418, as well as potentially useful agents such as bleomycin, showing that in principle it should be a generally useful marker for the production of transgenic rice plants (Caplan et al. 1992).

1.3 Novel Selectable Markers

1.3.1 Innocuous Markers

Public concern about the widespread use of antibiotic- and herbicide-resistance markers has, in some European countries, prompted the development of alternative marker systems that use innocuous selective regimens. Although such markers have yet to be tested in rice, they are worth mentioning here because it is only a matter of time before they are adopted. Currently, transgenic rice accounts for nearly 20% of genetically modified cropland worldwide, and the demands of consumers are likely to necessitate the use of such markers in the future commercialization of transgenics.

We describe two examples, both of which have been used successfully in other crop species. The first example is the *Escherichia coli* mannose phosphate isomerase gene (*manA*), which confers upon transformed cells the ability to catabolize mannose (Negrotto et al. 2000). To use this system, callus is regenerated on medium containing mannose as the sole carbon source, such that transformed cells grow normally, while the proliferation of nontransformed cells is restricted. The second example is the isopentyl transferase (*ipt*) gene from *Agrobacterium tumefaciens* (Ebinuma et al. 1997; Kunkell et al. 1999). The gene is located within the T-DNA of wild-type Ti plasmids and induces cytokinin synthesis in transformed plants. Based on its mutant phenotype, the gene was initially designated *tmr* for "tumor rooty" because its absence reduced the level of cytokinin in crown galls. As a marker, the functional *ipt* gene can be used to select plants on the basis of their ability to produce shoots from callus on medium lacking cytokinins.

1.3.2 Counterselectable Markers

The marker genes discussed so far are all *positive* selectable markers, meaning that the transformed plant survives in the *presence* of the marker, which confers *resistance* to (or allows growth on) medium containing an agent that is toxic to (or restricts the growth of) nontransformed cells. As an alternative to innocuous marker systems, some laboratories have explored strategies for removing marker genes from transgenic plants once the transgenic line is stable. This has required the development of *negative* selectable markers (also called counterselectable markers), meaning that plants are selected on the basis of the *absence* of the marker, which confers *sensitivity* to a specific agent. An example is the *tms2* gene, which is also found within the T-DNA of naturally occurring *Agrobacterium* Ti plasmids. The gene encodes indoleacetic acid hydrolase (IAAH), which converts naphthaleneacetamide (NAM) into naphthaleneacetic acid, a potent auxin. In the presence of exogenously applied NAM, transformed callus is unable to produce shoots due to the excess levels of auxin, therefore only callus lacking the gene is able to regenerate into full transgenic plants. This marker system has been used in tobacco and other model species, but has been evaluated in rice only recently (Uphadyaya et al. 2000).

1.4 Screenable Marker Genes

Screenable marker genes have a wide range of uses, either as independent genes or fusion constructs (Table 1.3). The earliest use of screenable markers in plants was to confirm transformation. Initially, the opine synthesis genes *nos* and *ocs* were utilized because these were already present on the T-DNA of wild-type Ti plasmids (reviewed by Dessaux and Petit 1994). However, heterologous marker genes such as *cat* (encoding chloramphenicol acetyltransferase) and *lacZ* (encoding β-galactosidase) were soon adopted for use in plants, due to

Table 1.3. Some uses for screenable marker genes in plants

Expression under the control of a constitutive promoter
- Confirming transformation (transient or stable)
- Determining transformation efficiency
- Labeling cells for lineage analysis
- Excision marker for transposable elements

Expression under the control of a cloned plant promoter
- Confirming expression in particular cell types, e.g. callus
- Determining spatiotemporal expression profile
- Investigating transcriptional induction in response to external signals
- Dissecting *cis*-acting regulatory elements

Expression as an uncharacterized transcriptional fusion
- Enhancer-trap constructs
- Gene-trap constructs

Expression as a fusion protein
- Rapid confirmation/quantification of recombinant protein expression
- Labeling and localizing of recombinant proteins
- Investigating protein trafficking in the cell
- Investigating the dynamics of intracellular structure
- Investigating protein transport at the whole-plant level, including the spread of viral infections

the availability of well-established and simple assay formats, and their applicability to many different species (Herrera-Estrella et al. 1983; Helmer et al. 1984). An important advantage of *lacZ* is that it can be used as a histological marker. The substrate X-gal is cleaved to yield an insoluble blue product that marks transformed cells brightly. However, many plant tissues have a high background of endogenous β-galactosidase activity, which can interfere with the interpretation of chimeric marker-gene activity. Therefore, in place of *cat* and *lacZ*, a number of more versatile screenable markers have been developed for use in plants. Three of these – *gusA* (β-glucuronidase), *luc* (luciferase) and *gfp* (green fluorescent protein) – have been widely used in rice, and are discussed below.

1.4.1 β-Glucuronidase (*gusA*)

The most widely used screenable marker in plants is the *E. coli* gene *gusA*, encoding the enzyme β-glucuronidase (GUS) (Jefferson et al. 1986). This marker is analogous to *lacZ*, and parallel assay formats are available, e.g. a fluorometric in vitro assay using the substrate MUG (4-methylumbelliferyl-β-D-glucuronide) and a colorimetric histological assay using the substrate X-gluc (5-bromo-4-chloro-3-indolyl-β-D-glucuronide). However, *gusA* has two major advantages over *lacZ* as a marker: (1) plants show minimal endogenous GUS activity, and (2) the gene itself is much smaller than *lacZ* and thus more versatile for vector development (Jefferson et al. 1986).

The *gusA* gene was the first screenable marker to be tested in rice (Christou et al. 1991). Immature embryos were bombarded with a plasmid containing *gusA* and the hygromycin resistance gene *hpt*. Transient GUS activity was apparent 24h after bombardment and was affected by many parameters including DNA amount, accelerating voltage, and carrier particle load. As independent genes under the control of a constitutive promoter, one of the most important uses of screenable markers in plants is to provide visual evidence of transformation and allow the evaluation of transformation efficiency in transient assays. In particle bombardment, for example, the conversion rate from transient to stable transformation is often as low as 1%, so the use of reporter-gene assays to establish parameters for robust transient expression is generally useful before going to the expense of attempting to regenerate transgenic plants. GUS has been widely used for this purpose, since transformation variables can influence both the number and size of GUS-positive regions in transformed explants, allowing the most appropriate bombardment parameters to be chosen. However, it should be emphasized that the optimal parameters for transient transformation may not be the same as those for stable transformation. The *gusA* gene was also used in many of the early attempts to generate transgenic rice plants using *Agrobacterium tumefaciens* (Raineri et al. 1990; Chan et al. 1992; 1993; Hiei et al. 1994; Dong et al. 1996; Rashid et al. 1996).

As well as its use as a marker to verify transformation, *gusA* has also been developed as a transcriptional fusion marker to investigate gene regulation. Rice is the most widely used cereal species for testing promoters as *gusA* reporter constructs (Izawa and Shimamoto 1996). Such experiments have led to the detailed characterization of the activities of commonly used constitutive promoters such as CaMV 35S, the rice actin-1 promoter and the maize ubiquitin-1 promoter (reviewed by McElroy and Brettell 1994). Furthermore, an increasing number of tissue-specific and developmentally regulated promoters have been studied by this method (comprehensively reviewed by Bilang et al. 1999). An example of GUS staining in rice is shown in Fig. 1.3.

In rice, the ability of *gusA* to report the activities of regulatory elements to which it becomes attached has also been exploited to identify and characterize new genes as part of a functional genomics strategy. Transposable elements and T-DNA have been widely used in plants as mutagens and tools to clone tagged genes (reviewed by Maes et al. 1999). The inclusion within such sequences of a promoterless marker gene such as *gusA* brings the marker under the control of any regulatory elements close to its integration site. In this way, the corresponding gene is not only tagged by the insertion, but its expression pattern can also be characterized. This allows screens for insertions into genes with specific expression patters, such as seed-specific or leaf-specific

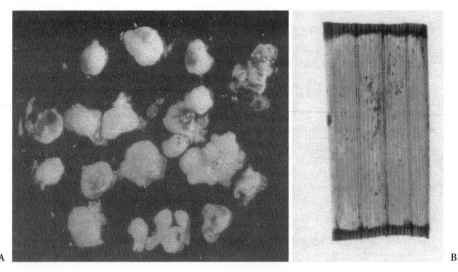

A   B

Fig. 1.3A,B. In situ glucuronidase (GUS) activity. **A** Rice callus several hours after bombardment, showing transient GUS activity that reports transformation efficiency. **B** Rice leaf section showing stable GUS activity. This leaf was isolated from a transgenic plant recovered after bombarding embryogenic callus with a *gusA* expression construct. (Courtesy of Dr. Pawan Agrawal, John Innes Centre)

genes (Topping et al. 1991; Sundaresan et al. 1995; Kertbundit et al. 1998; Campisi et al. 1999). This strategy has been used by Jeon et al. (2000) to generate over 18,000 transgenic rice lines transformed with T-DNA sequences containing a *gusA*-based gene-trap vector. The vector comprised a promoterless *gusA* gene downstream of a heterologous intron and several splice-acceptor sites. When this construct integrates within a rice gene, a fusion transcript is generated resulting in GUS activity that matches the expression pattern of the interrupted gene. The frequency of GUS activity in leaves, roots, flowers and seeds of transformed plants in this report ranged from 1.6 to 2%. It was possible to dissect these tissues and use the histological X-gluc assay to investigate the expression in specific cell types. For example, investigation of those lines with GUS activity in flowers revealed individual lines with marker-gene activity restricted to quite specific regions, such as the glumes, rachilla, stamens, carpels, lodicules, palea and lemma.

1.4.2 Firefly Luciferase (*luc*)

The luciferase (*luc*) gene, from the North American firefly *Photinus pyralis*, was first used as a screenable marker in plants by Ow et al. (1986). The enzyme catalyzes the oxidation of luciferin, in a reaction requiring oxygen, ATP and magnesium ions. In the presence of excess substrate, a flash of light is emitted that is proportional to the amount of enzyme. Important advantages of the luciferase system include its very high sensitivity and the rapid decay of the signal, allowing it to be used for the analysis of genes with rapidly oscillating expression profiles. In contrast, *gusA* produces a stable protein that persists in cells even after the promoter to which it is attached has been inactivated.

Luciferase has not been widely used in rice. It has been used for the visible screening of transformed rice protoplasts (Sadasivam and Gallie 1994), and transgenic callus and plants have been produced recently from immature embryos transformed with the *luc* gene by particle bombardment (Baruah-Wolff et al. 1999). A bacterial luciferase gene, *luxA*, has also been used as a marker in transgenic plants, although not in rice (Koncz et al. 1987).

1.4.3 Green Fluorescent Protein (*gfp*)

Green fluorescent protein (GFP) is a bioluminescent protein from the jellyfish *Aequoria victoria*. Over the last few years, GFP has emerged as an extremely versatile tool for the analysis of biological processes in plants and many other organisms (reviewed by Haseloff 1999; Naylor 1999). When exposed to blue or ultraviolet light, GFP emits bright green fluorescence that is proportional to the amount of protein present. However, unlike luciferase, GFP has no substrate requirements and can therefore be used as a vital marker to assay cellular processes in real time.

Initial experiments with GFP in plants had limited success due to the instability of expression (Haseloff and Amos 1995; Hu and Chen 1995; Niedz et al. 1995; Sheen et al. 1995). Modifications in terms of codon usage have been necessary for robust GFP expression in some plants (Chiu et al. 1996), and in *Arabidopsis*, the original *gfp* gene is expressed very poorly due to aberrant splicing. This problem has been addressed by removing a cryptic splice site (Haseloff et al. 1997). GFP has also been mutated to improve or modify various properties of the protein, including its excitation and emission wavelengths, signal intensity, and stability (e.g. Heim and Tsein 1996).

Green fluorescent protein was first evaluated as a marker in rice by Nagatani et al. (1997) who obtained transient expression in bombarded immature embryos. Transient GFP activity was also observed by Li and colleagues in rice callus less than 12 h after bombardment (Huang et al. 1997). The generation of transgenic rice plants expressing GFP was first reported by Vain et al. (1998), who used the re-engineered *mgfp4* gene to identify cells transformed by particle bombardment. The use of GFP in addition to traditional hygromycin selection had a number of advantages. The stably transformed tissue could be identified within 2 weeks of bombardment, rather than the 6–12 weeks required using antibiotic selection alone. Also, the system was uniformly applicable to all cultivars, whereas selection conditions typically differ from variety to variety. Recently, GFP has been used as an excision marker to track the activity of *Ac-Ds* elements in transgenic rice populations (Greco et al. 2001; Kohli et al. 2001; Fig. 1.4).

Fig. 1.4. Mosaic GFP activity in rice plants following somatic excision of the *Ac* transposable element, which contains the green fluorescent protein *gfp* gene controlled by the maize ubiquitin-1 promoter. (Kohli et al. 2001)

Although valuable as a marker for transformation, one of the major uses of GFP is as a fusion tag to track proteins in living cells (e.g. see Sheen et al. 1995; Kohler et al. 1997). In rice, translational fusions between GFP and a chloroplast-targeting signal have resulted in successful targeting of GFP fluorescence to chloroplasts (Jang et al. 1999). In this experiment, two constructs were introduced into rice embryos by *Agrobacterium*-mediated transformation, one in which the marker was expressed without a signal peptide under the control of the actin-1 promoter, and another in which it was expressed as a fusion protein using the rubisco small-subunit promoter. Transgenic rice plants were regenerated and analyzed by confocal microscopy. In those containing the control construct, GFP fluorescence was localized in the cytoplasm and nucleus. In those containing the fusion construct, GFP fluorescence was localized to the chloroplasts and non-green plastids. Furthermore, expression levels were 20 times higher in plants expressing the fusion construct, with GFP representing approximately 10% total soluble protein.

1.5 Strategies for Marker-Gene Delivery

Due to the requirement for marker genes, rice transformation necessarily involves the introduction of at least two genes into the target tissue at the same time – the transgene of interest and a screenable or selectable marker. Dual transformation can be achieved by either including both genes on the same vector, or using separate vectors, the latter process known as cotransformation.

For *Agrobacterium*-mediated transformation, the marker and experimental transgene are generally cloned in tandem on the same T-DNA. Although most binary vectors carry the marker adjacent to the right-border repeat, it is actually better for the marker to be placed at the left border, since T-DNA transfer has a right-to-left polarity and the left-border repeat is transferred to the plant cell last (Sheng and Citovsky 1996). With the marker close to the right-border repeat, there is an increased chance that interruption of the transfer process could generate transgenic plants that survive selection but carry the marker gene alone and not the transgene of interest. More recently constructed binary vectors have the marker gene at the left border (see Hellens and Mullineaux 2000). Co-delivery of two transgenes can also be accomplished by infecting plants with an *Argobacterium* strain carrying two different binary vectors, or by co-infecting plant tissue with two bacterial strains. It is not clear whether these strategies favor co-integration of the transgenes at the same locus, or independent integration events. The latter process could be favorable for the removal of marker genes from transgenic plant lines by sexual crossing (Komari et al. 1996). The type of Ti plasmid (octopine or nopaline) could play a more important role than the transformation strategy in determining transgene organization in the plant genome (Depicker et al. 1985; Jorgensen et al. 1987; De Block and Debrouwer 1991).

In direct DNA transfer methods, the selectable marker can be included with the transgene of interest on a "cointegrate vector" but this is not necessary, because the use of separate vectors also leads to a high frequency of co-transformation (Schocher et al. 1986; Christou and Swain 1990). In rice, the cointegrate vector and cotransformation strategies have been extensively compared using a variety of markers and genes of agronomic interest (e.g. Kohli et al. 1999; Gahakwa et al. 2000). In the vast majority of cases, both strategies result in the cointegration of all transgenes at a single locus. There have been occasional anecdotal reports of segregation between the marker and the transgene in some transgenic lines, but we must presume this is a very rare event.

Cotransformation can be advantageous, since it simplifies the cloning strategy required prior to transformation and removes any size constraints incumbent in the use of multiple genes. Chen et al. (1998) cotransformed rice with 14 separate plasmids containing various selectable and screenable marker genes, as well as genes of agronomic interest, the maximum reported to date, and succeeded in recovering transgenic plants in which 13 of these genes had co-integrated. In a recent, development, Fu et al. (2000) demonstrated the principle that rice can be co-transformed with multiple genes in the form of minimal linear expression cassettes lacking all vector backbone elements, and that the frequency of co-transformation is similar to that observed using traditional plasmid vectors.

References

Aldemita RR, Hodges TK (1996) *Agrobacterium tumefaciens*-mediated transformation of japonica and indica rice varieties. Planta 199:612–617

Ayres NM, Park WD (1994) Genetic transformation of rice. Crit Rev Plant Sci 13:219–239

Baruah-Wolff J, Harwood WA, Lonsdale DA, Harvey A, Hull R, Snape JW (1999) Luciferase as a reporter gene for transformation studies in rice (*Oryza sativa* L.). Plant Cell Rep 18:715–720

Bilang R, Futterer J, Sautter C (1999) Transformation of cereals. Genet Eng 21:113–157

Campisi L, Yang Y, Yi Y, Hellig E, Herman B, Cassista AJ, Allen DW, Xiang H, Jack T (1999) Generation of enhancer trap lines in *Arabidopsis* and characterization of expression patterns in the inflorescence. Plant J 17:699–707

Cao J, Duan X, McElroy D, Wu R (1992) Regeneration of herbicide-resistant transgenic rice plants following microprojectile-mediated transformation of suspension culture cells. Plant Cell Rep 11:586–591

Caplan A, Dekeyser R, van Montagu M (1992) Selectable markers for rice transformation. Methods Enzymol 216:426–441

Chan M-T, Lee T-M, Chang H-H (1992) Transformation of indica rice (*Oryza sativa* L.) mediated by *Agrobacterium tumefaciens*. Plant Cell Physiol 33:577–583

Chan M-T, Chang H-H, Ho S-L, Tong W-F, Yu S-M (1993) *Agrobacterium*-mediated production of transgenic rice plants expressing a chimeric α-amylase promoter/β-glucuronidase gene. Plant Mol Biol 22:491–506

Chen LL, Marmey P, Taylor NJ, Brizard JP, Espinoza C, D'Cruz P, Huet H, Zhang SP, de Kochko A, Beachy RN, Fauquet CM (1998) Expression and inheritance of multiple transgenes in rice plants. Nat Biotechnol 16:1060–1064

Chiu W, Niwa Y, Zeng W, Hirano T, Kobayashi H, Sheen J (1996) Engineered GFP as a vital reporter in plants. Curr Biol 3:325–330

Christou P, Ford TL (1995) Parameters influencing stable transformation of rice embryonic tissue and recovery of transgenic plants using electric discharge particle acceleration. Ann Bot 75: 407–413

Christou P, Swain WF (1990) Cotransformation frequencies of foreign genes in soybean cell cultures. Theor Appl Genet 90:97–104

Christou P, Ford TL, Kofron M (1991) Production of transgenic rice (Oryza sativa L.) from agronomically important indica and japonica varieties via electric discharge particle acceleration of exogenous DNA into immature zygotic embryos. Bio/technology 9:957–962

Dale EC, Ow DW (1991) Gene transfer with subsequent removal of the selection gene from the host genome. Proc Natl Acad Sci USA 88:10558–10562

Datta SK, Peterhans A, Datta K, Potrykus I (1990) Genetically engineered fertile indica rice recovered from protoplasts. Bio/technology 8:736–740

De Block M, Debrouwer D (1991) Two T-DNAs co-transformed into Brassica napus by a double Agrobacterium tumefaciens infection are mainly integrated at the same locus. Theor Appl Genet 82:257–263

De Block M, Botterman J, Vandewiele M, Dockx J, Thoen C, Gossele V, Rao Movva N, Thompson C, van Montagu M, Leemans J (1987) Engineering herbicide resistance in plants by expression of a detoxifying enzyme. EMBO J 6:2513–2518

Dekeyser R, Claes B, Marichal M, van Montagu M, Caplan A (1989) Evaluation of selectable markers for rice transformation. Plant Physiol 90:217–223

Depicker A, Herman L, Jacobs A, Schell J, van Montagu M (1985) Frequencies of simultaneous transformation with different T-DNAs and their relevance to the Agrobacterium/plant cell interaction. Mol Gen Genet 201:477–484

Dessaux Y, Petit A (1994) Opines as screenable markers for plant transformation. In: Gelvin SB, Schilperoort RA (eds) Plant Molecular Biology Manual, 2nd edn. Marcel Dekker, New York, pp 1–12

Dong J, Teng W, Buchholz WG, Hall TC (1996) Agrobacterium-mediated transformation of javanica rice. Mol Breed 2:267–276

Ebinuma H, Sugita K, Matsunaga E, Yamakado M (1997) Selection of marker-free transgenic plants using the isopentenyl transferase gene. Proc Natl Acad Sci USA 94:2117–2121

Fu X, Duc LT, Fontana S, Bong BB, Tinjuangjun P, Sudhakar D, Twyman RM, Christou P, Kohli A (2000) Linear transgene constructs lacking vector backbone sequences generate low-copy-number transgenic plants with simple integration patterns. Transgenic Res 9:11–19

Gahakwa D, Bano Maqbool S, Fu X, Sudhakar D, Christou P, Kohli A (2000) Transgenic rice as a system to study the stability of transgene expression: multiple heterologous transgenes show similar behaviour in diverse genetic backgrounds. Theor Appl Genet 101:388–399

Goldsbrough AP, Lastrella CN, Yoder JI (1993) Transposition mediated repositioning and subsequent elimination of marker genes from transgenic tomato. Bio/technology 11:1286–1292

Golovkin MV, Abraham M, Morocz S, Bottka S, Feher A, Dudits D (1993) Production of transgenic maize plants by direct DNA uptake into embryogenic protoplasts. Plant Sci 90:41–52

Greco R, Ouwerkerk PBF, Anke JCT, Favalli C, Beguiristain T, Puigdomenech P, Colombo L, Hoge JHC, Pereira A (2001) Early and multiple Ac transpositions in rice generated by an adjacent strong enhancer. Plant Mol Biol 46:215–227

Haseloff J (1999) GFP variants for multispectral imaging of living cells. Methods Cell Biol 58: 139–166

Haseloff J, Amos B (1995) GFP in plants. Trends Genet 11:328–329

Haseloff J, Siemering KR, Prasher DC, Hodge S (1997) Removal of a cryptic intron and subcellular localization of green fluorescent protein are required to mark transgenic Arabidopsis plants brightly. Proc Natl Acad Sci USA 94:2122–2127

Heim R, Tsein RY (1996) Engineering green fluorescent protein for improved brightness, longer wavelengths and fluorescence resonance energy transfer. Curr Biol 1996 6:178–182

Hellens R, Mullineaux P (2000) A guide to *Agrobacterium* binary Ti vectors. Trends Plant Sci 5: 446–451

Helmer G, Casadaban M, Bevan M, Kayes L, Chilton M-D (1984) A new chimeric gene as a marker for plant transformation: the expression of *Escherichia coli* β–galactosidase in sunflower and tobacco cells. Bio/technology 2:520–527

Herrera-Estrella L, Depicker A, van Montagu M, Schell J (1983) Expression of chimeric genes transferred into plant cells using a Ti-plasmid-derived vector. Nature 303:209–213

Hiei Y, Ohta S, Komari T, Kumashiro T (1994) Efficient transformation of rice (*Oryza staiva* L.) mediated by *Agrobacterium* and sequence analysis of the T-DNA. Plant J 6:271–282

Hu W, Chen C (1995) Expression of *Aequoria* green fluorescent protein in plant cells. FEBS Lett 369:331–334

Huang Y, Xu XP, Li BJ (1997) Improved green fluorescent protein as a fast reporter of gene expression in plant cells. Biotechnol Tech 11:133–136

Izawa T, Shimamoto K (1996) Becoming a model plant: the importance of rice to plant science. Trends Plant Sci 1:95–99

Jang IC, Nahm BH, Kim JK (1999) Subcellular targeting of green fluorescent protein to plastids in transgenic rice plants provides a high-level expression system. Mol Breed 5:453–461

Jefferson RA, Kavanagh TA, Bevan M (1987) GUS fusions: β-glucuronidase as a sensitive and versatile gene fusion marker in higher plants. EMBO J 6:3910–3907

Jeon J-S, Lee S, Jung K-H, Jun S-H, Jeong D-H, Lee J, Kim C, Jang S, Lee S, Yang K, Nam J, An K, Han M-J, Sung R-J, Choi H-S, Yu J-H, Choi J-H, Cjo S-Y, Cha S-S, Kim S-I, An G (2000) T-DNA insertional mutagenesis for functional genomics in rice. Plant J 22:561–570

Jorgensen RA, Snyder C, Jones JDG (1987) T-DNA is organized predominantly in inverted repeat structures in plants transformed with *Agrobacterium tumefaciens* C58 derivatives. Mol Gen Genet 207:471–477

Kertbundit S, Linacero R, Rouze P, Galis I, Macas J, Deboeck F, Renckens S, Hernalsteens J-P, de Greve H (1998) Analysis of T-DNA-mediated translational β-glucuronidase gene fusions. Plant Mol Biol 36:205–217

Kohler RH, Zipfel WR, Webb WW, Hanson M (1997) The green fluorescent protein as a marker to visualize plant mitochondria in vivo. Plant J 11:613–621

Kohli A, Gahakwa D, Vain P, Laurie DA, Christou P (1999) Transgene expression in rice engineered through particle bombardment: molecular factors controlling stable expression and transgene silencing. Planta 208:88–97

Kohli A, Xiong J, Greco R, Christou P, Pereira A (2001) Tagged Transcriptome Display (TTD) in *Indica* rice using *Ac* transposition. Mol Gen Genet 266:1–11

Komari T, Hiei Y, Saito Y, Murai N, Kumashiro T (1996) Vectors carrying two separate T-DNAs for co-transformation of higher plants mediated by *Agrobacterium tumefaciens* and segregation of transformants free from selection markers. Plant J 10:165–174

Koncz C, Olsson O, Langridge WHR, Schell J, Szalay AA (1987) Expression and assembly of functional bacterial luciferase in plants. Proc Natl Acad Sci USA 84:131–135

Kunkell T, Niu QW, Chan YS, Chua NH (1999) Inducible isopentenyl transferase as a high-efficiency marker for plant transformation. Nat Biotechnol 17:916–919

Li ZJ, Hayashimoto A, Murai N (1992) A sulfonylurea herbicide resistance gene from *Arabidopsis thaliana* as a new selectable marker for production of fertile transgenic rice plants. Plant Physiol 100:662–668

Maes T, de Keukeleire P, Gerats T (1999) Plant tagnology. Trends Plant Sci 4:90–96

McElroy D, Brettell RIS (1994) Foreign gene expression in transgenic cereals. Trends Biotechnol 12:62–68

Meijer EGM, Schilperoort RA, Rueb S, van Osruygrok PE, Hensgens LAM (1991) Transgenic rice cell lines and plants – expression of transferred chimeric genes. Plant Mol Biol 16:807–820

Nagatani N, Takumi S, Tomiyama M, Shimada T, Tamiya E (1997) Semi-real time imaging of the expression of a maize polyubiquitin promoter-GFP gene in transgenic rice. Plant Sci 124:49–56

Naylor LH (1999) Reporter gene techniques: the future is bright. Biochem Pharmacol 58:745–757

Negrotto D, Jolley M, Beer S, Wenck AR, Hansen G (2000) The use of phosphomannose-isomerase as a selectable marker to recover transgenic maize plants (*Zea mays* L.) via *Agrobacterium* transformation. Plant Cell Rep 19:798–803

Niedz RP, Sussman MR, Satterlee JS (1995) Green fluorescent protein: an in vitro reporter of plant gene expression. Plant Cell Rep 14:403–406

Oard JH, Linscombe SD, Braverman MP, Jodari F, Blouin DC, Leech M, Kohli A, Vain P, Cooley JC, Christou P (1996) Development, field evaluation, and agronomic performance of transgenic herbicide resistant rice. Mol Breed 2:359–368

Ow DW, Wood KV, DeLuca M, de Wet JR, Helinski DR, Howell SH (1986) Transient and stable expression of the firefly luciferase gene in plant cells and transgenic plants. Science 234: 856–859

Park SH, Pinson SRM, Smith RH (1996) T-DNA integration into genomic DNA of rice following *Agrobacterium* inoculation of isolated shoot apices. Plant Mol Biol 32:1135–1148

Raineri DM, Bottino P, Gordon MP, Nester EW (1990) *Agrobacterium*-mediated transformation of rice (*Oryza sativa* L.). Bio/technology 8:33–38

Rashid H, Yoki S, Toriyama K, Hinata K (1996) Transgenic plant production mediated by *Agrobacterium* in indica rice. Plant Cell Rep 15:727–730

Rathore KS, Chowdhury VK, Hodges TK (1993) Use of *bar* as a selectable marker gene for the production of herbicide-resistant rice plants from protoplasts. Plant Mol Biol 21:871–884

Sadasivam S, Gallie DR (1994) Isolation and transformation of rice aleurone protoplasts. Plant Cell Rep 13:394–396

Schocher RJ, Shillito RD, Saul RD, Paszkowski SJ, Potrykus I (1986) Co-transformation of unlinked foreign genes into plants by direct gene transfer. Bio/technology 4:1093–1096

Sheen J, Hwang S, Niwan Y, Kobayashi H, Galbraith DW (1995) Green fluorescent protein as a new vital marker in plant cells. Plant J 8:777–784

Sheng OJ, Citovsky V (1996) *Agrobacterium*-plant cell DNA transport: have virulence proteins, will travel. Plant Cell 8:1699–1710

Sudhakar D, Duc LT, Bong BB, Tinjuangjun P, Bano Maqbool S, Valdez M, Jefferson R, Christou P (1998) An efficient rice transformation system utilizing mature seed-derived explants and a portable, inexpensive particle bombardment device. Transgenic Res 7:289–294

Sundaresan V, Springer P, Volpe T, Haward S, Jones JDG, Dean C, Ma H, Martienssen RA (1995) Patterns of gene action in plant development revealed by enhancer trap and gene trap transposable elements. Genes Dev 9:1797–1810

Topping JF, Wei W, Lindsey K (1991) Functional tagging of regulatory elements in the plant genome. Development 112:1009–1019

Toriyama K, Arimoto Y, Uchimiya H, Hinata K (1988) Transgenic plants after direct gene transfer into protoplasts. Bio/technology 6:1072–1074

Twyman RM, Christou P, Stöger E (2001) Genetic transformation of plants and their cells. In: Oksman-Caldentey KM, Barz W (eds) Plant Biotechnology and Transgenic Plants. Marcel Dekker, New York

Upadhyaya NM, Zhou XR, Wu LM, Ramm K, Dennis ES (2000) The *tms2* gene as a negative selection marker in rice. Plant Mol Biol Rep 18:227–233

Vain P, Worland B, Kohli A, Snape JW, Christou P (1998) The green fluorescent protein (GFP) as a vital screenable marker in rice transformation. Theor Appl Genet 96:164–169

Van den Elzen PJM, Townsend J, Lee KY, Bedbrook JR (1985) A chimeric hygromycin resistance gene as a selectable marker in plant cells. Plant Mol Biol 5:299–302

Zhang HM, Yang H, Rech EL, Gold TJ, Davis AS, Mulligan BJ, Cocking EC, Davey MR (1988) Transgenic rice plants produced by electroporation-mediated plasmid uptake into protoplasts. Plant Cell Rep 7:379–383

Zhang W, Wu R (1988) Efficient regeneration of transgenic plants from rice protoplasts and correctly regulated expression of the foreign gene in the plants. Theor Appl Genet 76:835–840

Zubko E, Scutt C, Meyer P (2000) Intrachromosomal recombination between *attP* regions as a tool to remove selectable marker genes from tobacco transgenes. Nat Biotechnol 18:442–445

2 Use of Green Fluorescent Protein to Detect Transformed Shoots

J. MOLINIER and G. HAHNE

2.1 Introduction

Visible reporter genes are a very useful approach to evaluating the efficiency of newly developed protocols in plant transformation technology. The basic idea is to distinguish transformed regenerants (shoots or somatic embryos) in an heterogeneous population from untransformed ones without a reduction in the viability of the regeneration tissue, as is the case with selection markers (i.e. antibiotics, herbicides, etc.). Until recently, the reporter genes used for this approach were β-glucuronidase (*uidA*) and luciferase (*Luc*) (Ow et al. 1986; Jefferson et al. 1987). Both of these detection procedures require exogenous substrates to visualize the enzymatic activity of the gene products, which in the case of β-glucuronidase (GUS) is destructive for tissues because of the toxicity of the substrate (Jefferson et al. 1987). Furthermore, fixation of the tissue is often required for histochemical detection assays. Detection of luciferase (LUC) expression is possible in vivo, but an exogenous substrate, luciferin, must be applied. Furthermore, the light level generated by the luciferin/luciferase system is low, and a powerful and expensive low-light detection camera is needed for in vivo detection. Alternatively, enzyme activity can be measured in crude extracts for both marker systems.

GUS and LUC have been extensively used, and most of the available transformation protocols have been developed based on one of these two genes. The GUS method is easy and inexpensive, but GUS is, in most cases, not a vital stain, whereas LUC allows observation of living tissue or even entire plants but requires expensive and sophisticated equipment. The ideal reporter gene would thus combine in vivo monitoring with a minimum of necessary equipment.

Green fluorescent protein (GFP), from the jellyfish *Aequorea victoria*, emits bright green fluorescence upon excitation with near-UV or blue light and is visible under normal fluorescent room light (Morise et al. 1974). After modification of the coding sequence in order to optimize the fluorescence properties, GFP has been extensively used as reporter gene in heterologous systems (Prasher 1995). GFP is a small protein that does not require the presence of additional factors other than oxygen for its detection, and it is compatible with transcriptional and translational fusions with other proteins. GFP has been expressed in plants (Haseloff and Amos 1995; Haseloff and Siemering 1998), in which it has been used to study the expression patterns of pro-

Molecular Methods of Plant Analysis, Vol. 22
Testing for Genetic Manipulation in Plants
Edited by J.F. Jackson, H.F. Linskens, and R.B. Inman
© Springer-Verlag Berlin Heidelberg 2002

moters (Sheen et al. 1995; Nagatani et al. 1997) and to follow short- and long-distance movements of proteins and viruses (Itaya et al. 1997). There are a few reports available concerning the technical aspects of GFP detection and, more specifically, the use of GFP for the selection of transgenic explants.

This chapter will focus on the use of GFP as a selectable marker for the detection of transgenic shoots. Particular attention will be given to describing the characteristics of the GFP protein and its potential to be used in transformation protocols. The detection equipment required as well as frequently encountered problems and guidelines for troubleshooting will also be reviewed here.

2.2 GFP: Suitable as a Visually Selectable Marker *In Planta*?

2.2.1 Important Properties of the Protein

Native GFP has two excitation maxima and hence emits visible light at 509 nm when excited by blue or near-UV light (395/475 nm), without the need for additional, exogenous substrates or cofactors other than oxygen (Prasher 1995). This allows GFP to be detected in vivo without damaging cells or tissues using classical equipment for the detection of fluorescence, e.g. a UV hand-lamp or a fluorescence microscope.

Modification of the primary structure of the protein has resulted in improved spectral and physical features compared to the original GFP. The modified versions of GFP are more efficiently excited by visible light, emit brighter fluorescence of varying wavelengths and are more stable in heterologous systems, such as bacteria, *Caenorhabditis elegans*, *Drosophila* and plants (Chalfie et al. 1994; Inouye and Tsuji 1994; Wang and Hazelrigg 1994; Davis and Viestra 1998). The mutations mainly concern preferences in codon usage by different organisms (e.g. increases in GC content for use in higher eukaryotes), the chromophore (amino acids 65 and/or 66) and the part of the protein involved in folding efficiency and stability. Currently available and frequently used GFP mutants have their maximal excitation peak at 475–490 nm and a maximal emission peak around 509 nm (for review, see Wintz 1999), although proteins with emission peaks at 447 nm (BFP, blue), 480 nm (CFP, turquoise) and 527 nm (YFP, yellow) have became available recently. While useful in double-detection experiments, their lower light emission make them less useful as visual in vivo markers. Nonetheless, these GFPs can be efficiently visualized in vivo using a UV lamp, an epifluorescent dissecting microscope or a fluorescent microscope. The necessary amount of GFP protein for the easy and unambiguous detection of fluorescence in living cells is very high compared to other reporter proteins currently used (e.g. GUS, LUC). Thus, GFP expression must be high in order to ensure that a sufficient number of molecules are accumulated. At least 10,000 GFP molecules per cell are necessary for the signal to be detectable, whereas around 100 GUS or LUC molecules are necessary due to enzymatic amplification of the detection signal. This is important when choosing the promoter that will be fused to the GFP-encoding sequence. It is also important to take into consideration the amount of time necessary for accumulation of the protein, its correct folding and its oxidation to occur.

2.2.2 Properties of a Useful Selectable Marker in Plant Transformation Technology

In plant transformation technology, marker genes such as *nptII* and *bar* are frequently used to select the transformed tissue population. Visually detectable markers such as GUS and LUC are mostly employed during subsequent improvement of the transformation protocol. For this purpose, the reporter gene should accurately detect transformed shoots or sectors of organs from untransformed ones without interfering with the regeneration process. Ideally, expression would be monitored in vivo and detection would require minimal equipment, compatible with handling the plant material under sterile conditions. A useful selectable marker should be independent of the type of organism or of a specific metabolic pathway. GFP seems to satisfy most of these requirements. GFP detection does not require exogenous substrate since it uses only oxygen, which is readily available at the sites where GFP protein accumulates. Moreover, its expression is easily monitored in vivo using either hand-held UV lamps (Morise 1974; Pang et al. 1996), epifluorescent dissecting microscopes, or epifluorescent microscopes (Pang et al. 1996). In addition, GFP expression and detection does not interfere with regeneration in different species (rice, oat, sunflower, tobacco, barley, lettuce, sugarcane, etc.) even though some decrease in regeneration efficiency has been reported in *Arabidopsis* (Haseloff et al. 1997).

However, certain plant tissues set natural limits to the use of GFP as a visual reporter because they contain large amounts of fluorescent molecules, such as chlorophyll and phenolic compounds. GFP detection may become problematic in such cases because the distinction of GFP-induced fluorescence from endogenous fluorescence may be difficult, even impossible, due to quenching. Careful troubleshooting and the choice of appropriate counter-measures may allow this problem to be overcome in many cases. Possible solutions are discussed below.

2.3 GFP Expression and Detection in Primary Transformed Tissues

2.3.1 Transient Expression and GFP Detection

GFP encompasses the main features required for a visually selectable marker to be useful in the improvement of transformation protocols. The development of a new protocol usually starts by evaluating transient GFP expression following biolistic (particle bombardment) or *Agrobacterium*-mediated gene transfer. Transient expression experiments yield information about which cell types are competent for DNA integration and regeneration. They also allow determination of expression levels and optimization of the detection procedure. In barley, sugarcane, maize, oat and citrus tissues, GFP was found to be easily detectable using an epifluorecent dissecting microscope or a fluorescent microscope 12 h after bombardment or coculture (Ahlandsberg et al. 1999; Elliott et al. 1999; Ghorbel et al. 1999; Kaeppler et al. 2000). In contrast, Vain et al. (1998) observed very low level of GFP fluorescence (mGFP4, Haseloff and Amos 1995) in rice tissues during the first 10 days following bombardment. It should be noted that the time required for the chromophore oxidation process is 4 h in wild-type GFP. This process and the accumulation of the protein are important points to consider for the detection procedure. In all the experiments, the GFP coding sequence was fused to a strong constitutive promoter (35S CaMV, Act1, Ubi1, etc.) in order to obtain the required high expression level. For successful production of transgenic plants, it is necessary for the transformed cells and for the regenerating cells to coincide. This is frequently a function of the experimental conditions (explant choice, transformation method, media composition, culture conditions, etc.). GFP allows optimization of these parameters without the necessity for the analyzed tissue to be sacrificed. A well-adapted procedure and proper equipment for the detection of GFP fluorescence are critical for the success of this optimization step.

2.3.2 Detection Equipment and Troubleshooting

Transient GFP expression is based on detection of the protein in the early stages (i.e. 12 h to 3 days) after bombardment or coculture; both of these procedures are known to result in damage to the plant tissues. Fluorescence is frequently observed already in healthy plant tissues but accumulates particularly as a result of early wound reactions. This accumulation may complicate GFP detection due to partial or complete overlap of the excitation and/or emission peaks. In tobacco leaf discs, 3 days after coculture with *Agrobacteria*, a mix of yellow and green fluorescence was observed under the epifluorescence dissecting microscope (Fig. 2.1A). Autofluorescence with spectral characteristics similar to GFP does occur in this tissue, which made unambiguous distinction of the marker gene difficult (Molinier et al. 2000). Vain et al. (1998) observed transient GFP expression in immature zygotic embryos of rice with the same difficulties and described the emission of pale-yellow to orange fluorescence from wounded and necrotic tissues. In the case of green tissues, red chlorophyll autofluorescence overlaps with GFP fluorescence and results in partial quenching, thus significantly decreasing its detection level. A judicious choice or adaptation of the detection procedure is therefore absolutely required for unambiguous detection of GFP fluorescence. In most of the experiments using GFP as a marker for transgenic tissue, fluorescent microscopes or epifluorescent dissecting microscopes are used, depending on the sample size. Epifluorescent microscopes are equipped with powerful mercury lamps and a large choice of excitation and emission filters, allowing the choice between band-pass or long-pass emission filters. Band-pass filters allow the emitted wave length to be more specifically detected, but they suffer from reduced transmitted-energy levels compared to the long-pass filters. Unfortunately, neither upright nor inverted fluorescence microscopes are well-adapted to working in aseptic conditions with in-vitro-cultured plant tissue. The use of a dissecting epifluorescent microscope is preferable in this case; however, early models were not designed for use with plants, as they feature a fix-mounted long-pass filter that does not cut off the chlorophyll autofluorescence (Fig. 2.1F). More recent models incorporate filters that are readily adaptable to plant work. Their band-pass filters (460–500/505/510–560 nm) eliminate the background of chlorophyll fluorescence (Fig. 2.1C, G) which was so problematic with the previous filter sets (Elliott et al. 1999; Molinier et al. 2000, see website: http//IBMP.u-strasbg.fr/dep_div_cell/Hahne/GFPHahne.html).

In transient experiments, after biolistic gene transfer, GFP expression can easily be monitored because explants synthesize less autofluorescing phenolic compounds that interfere with detection of the GFP signal. Callus or cell suspensions are ideally suited to such experiments because they are usually devoid of chlorophyll. In transient expression experiments involving *Agrobacterium*, GFP should be used cautiously due to the defense reaction of the plant tissue and the large amount of fluorescent molecules emerging from infected and wounded zones which perturb the detection of GFP fluorescence (Fig. 2.1A).

2.3.3 Stable Expression and GFP Detection in Primary Transformed Tissues

Rapid and noninvasive detection of stably transformed shoots or somatic embryos is possible with GFP-based selection. Table 2.1 gives some examples of the GFPs and the respective detection procedures used in several transformation protocolsand the detection procedures performed. In citrus, Ghorbel et al. (1999) succeeded in very early selection of green fluorescent calli emerging from cocultured internodal stem segments. GFP fluorescence was monitored with an epifluorescent dissecting microscope equipped with a long-pass filter, allowing the explants to be handled under aseptic conditions and decreasing the number of escapes (nontransgenic shoots) regenerating from

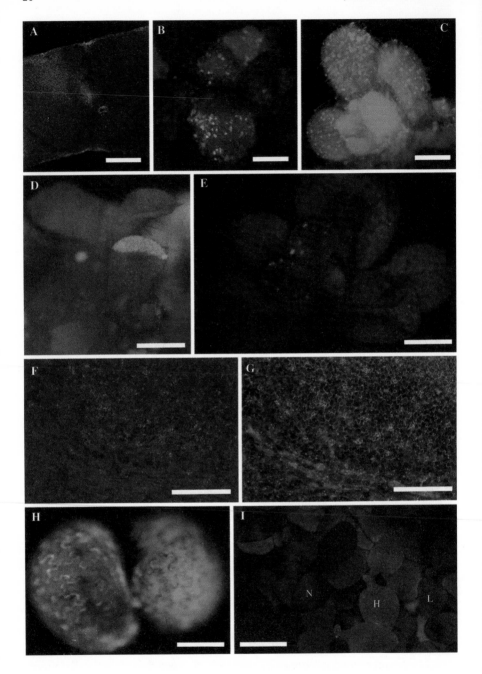

untransformed calli. Elliott et al. (1999) followed a similar approach in sugar-cane, maize and lettuce, following GFP fluorescence in clusters of cells every 3 days after transformation. They observed that most of the brightest clusters formed calli and transgenic explants (Fig. 2.1B, D). This approach is very useful for the selection of independent transformation events. Elliott et al. (1999) also compared several constructs for their potential in such experiments, i.e. sGFPS65T (Chiu et al. 1996) and mGFP5′ER (Haseloff et al. 1997), under the control of constitutive promoters (Ubi-1 or 35S CaMV), for selecting trans-genic tobacco, sugarcane or lettuce. They observed that GFP emitted brighter fluorescence with the sGFPS65T construct than with the mGFP5′ER construct, which is targeted to the endoplasmic reticulum. This observation is not sur-prising because the location of GFP in the cell (cytoplasm, organelles, nucleus) influences the final amount of protein that accumulates. Plant cells do not contain a large amount of cytoplasm which, due to the large vacuole, is dis-tributed thinly on the periphery of the cell. This is the reason why in young tissue, with smaller vacuoles and relatively more cytoplasm, GFP is more readily detected than in older tissue with more elongated cells (Fig. 2.1E). sGFPS65T localizes in the nucleus and cytoplasm and is currently the most frequently used GFP type in plant transformation protocols (Table 2.1). Nevertheless different GFPs are available for specific uses and can be adapted to plant species and/or detection procedures.

In most of the described transformation experiments, the regeneration process was indirect (indirect morphogenesis or somatic embryogenesis), for example, with *Arabidopsis*, tobacco and wheat, there was an intermediate callus phase. Indirect regeneration protocols are preferable because they allow the GFP-positive cell mass to be easily screened for, so that only the interest-ing (GFP-positive) callus pieces are subcultured. By using this method Vain et al. (1998) were able to significantly enhanced the efficiency of rice

◄———————————————————————————————————————

Fig. 2.1A–I. GFP detection (sGFPS65T) during different steps of *Agrobacterium*-mediated trans-formation of tobacco leaf fragments. A Transient GFP expression detected in a leaf fragment 3 days after coculture. Note the overlap of the yellow and green fluorescences in the wounded zones; *bar* 5 mm. B GFP-positive calli observed under the epifluorescence dissecting microscope with a long-pass filter (Leica MZ12 460–500/510 nm); *bar* 5 mm. C GFP-positive shoot observed under band-pass filter (460–500/505/ 510–560 nm); *bar* 5 mm. D Different GFP expression patterns in a regenerant population observed under the epifluorescence dissecting microscope; *bar* 5 mm. E A 6-week-old regenerant from a GFP-positive callus observed under the epifluorescence dissecting microscope. Note that only younger leaves are GFP-positive; *bar* 1 cm. F GFP-expressing leaf (2-week-old) observed under the upright microscope using a long-pass filter combination (460–500/505/ 510 nm); *bar* 500 µm. G GFP-expressing leaf (2-week-old) observed under the upright microscope using a band-pass filter combination (460–500/505/ 510–560 nm); *bar* 500 µm. H Wild-type (*left*) and GFP-expressing (*right*) seeds observed under the epifluorescence dissecting microscope; *bar* 1 mm. I Eight-day-old seedlings. Note the three fluorescence classes with high (*H*), low (*L*) and no (*N*) fluorescence corresponding to segregation events (homozy-gous, hemizygous and wild-type plants); *bar* 1 mm

Table 2.1. Examples of transformation experiments using GFP as the visual marker gene. *BP* Band-pass filter (cuts off chlorophyll-red autofluorescence), *CM* confocal microscope, *DM* dissecting microscope, *Ex* excitation, *Em* emission, *FM* fluorescent microscope, *IM* indirect morphogenesis, *ISE* indirect somatic embryogenesis, *UVHL* UV hand lamp

Plant	GFP type	Transformation technique	Regeneration process	GFP detection	Filter set	References
Apple	sGFPS65T	*Agrobacterium*	IM	DM	Ex: 450–490 nm Em: 515 nm or BP filter	Maximova et al. (1998)
		Agrobacterium	IM	DM		
Arabidopsis	mGFP5'ER	*Agrobacterium*	IM	FM	Ex: 355–425 nm Em: 460 nm	Haseloff et al. (1997); Niwa et al. (1999)
	mtsGFP65T			CM	Ex: 488 nm Em: 515–545 nm	
Arabidopsis, tobacco, maize, wheat	pGFPintron	Biolistic	IM/ISE	DM	Ex: 460–500 nm Em: 510–560nm	Pang et al. (1996)
	sGFPS65Tintron	*Agrobacterium*		UV HL		
Citrus	sGFPS65T	*Agrobacterium*	IM	DM	Ex: 460–500 nm Em: 505 nm	Ghorbel et al. (1999)
Lettuce, sugarcane, tobacco, maize	sGFPS65T	Biolistic	IM/ISE	DM	Ex: 460–500 nm Em: 510 nm or BP filter	Elliott et al. (1999)
	mGFP5'ER	*Agrobacterium*				
Maize	58 mutations	Biolistic	ISE	FM	Ex: 375–405 nm Em: 480–520 nm	Van der Geest and Petiolino (1998)
Oat	sGFPS65T	Biolistic	ISE	DM	Ex: 450–490 nm Em: 515–555 nm	Kaeppler et al. (2000)
Rice	mGFP4	Biolistic	ISE	DM UV HL	Ex: 460–500 nm Em: 510 nm	Vain et al. (1998)
Wheat	sGFPS65T	Biolistic	IM	DM	Ex: 450–490 nm Em: 500–550 nm	Jordan (2000)

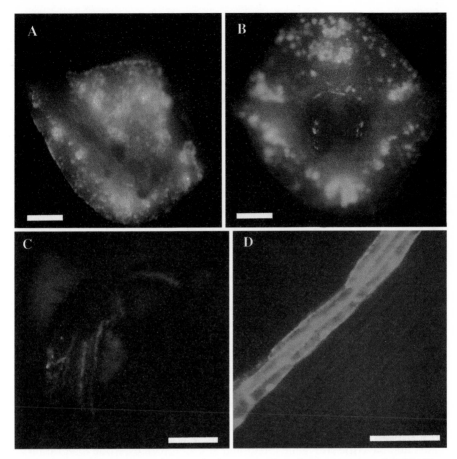

Fig. 2.2A–D. *Agrobacterium*-mediated transformation of sunflower using GFP (sGFPS65T) as visible marker gene. **A** Transient GFP expression in embryo axe 1 day after coculture; *bar* 1 mm. **B** Transient GFP expression in embryo axe 3 days after coculture. Note the red chlorophyll autofluorescence progressively overlapping the GFP fluorescence; *bar* 1 mm. **C** A 5-week-old GFP-positive shoot; *bar* 5 mm. **D** GFP-positive sunflower leaf (6-week-old) observed under band-pass filter (460–500/505/ 510–560 nm); *bar* 500 µm

transformation, thus reducing the quantity of handled tissue. To date no protocol has been published demonstrating that GFP-based selection could be used in transformation experiments involving a direct regeneration process. Our experience with transformation of sunflower is that GFP could not be observed in regenerated shoots before the fifth week after coculture (Fig. 2.2C, D), although transient GFP expression is easily detectable the first day after coculture (Fig. 2.2A) before it is progressively overlapped by the chlorophyll red fluorescence in the following 3 days (Fig. 2.2B) before disappearing (not shown). The GFP fluorescence then becomes undetectable until the fifth week in all cocultered and regenerated tissue. This might be due to the levels of expression and detection of GFP in the regenerant, green tissue. These results suggest that it is difficult to select early transformation events through this regeneration procedure (Molinier, unpublished results). Etiolation of the regenerating shoots, when possible, may alleviate the problem without interfering with the regeneration process.

2.4 GFP for Screening of Segregating Populations

It has been reported that the intensity of the emitted light is a function of the quantity of the accumulated protein (Blumenthal et al. 1999). This means that populations consisting of individuals with different GFP content can be screened based on the GFP fluorescence intensity (Table 2.2). In tobacco, GFP-positive seeds are easily distinguished from wild-type seeds only a few hours after imbibition (Fig. 2.1H; Molinier et al. 2000). Van der Geest and Petiolino (1998) observed bright GFP fluorescence in 50% of maize pollen and concluded that this observation was consistent with Mendelian segregation of a single dominant locus. In oat and wheat, GFP-based segregation ratios were determined in the T1 generation (Jordan 2000; Kaeppler et al. 2000). Leffel et al. (1997) observed classes of different fluorescence in their plants and could correlate the fluorescence level with the amount of GFP, suggesting that this distribution fits Mendelian segregation. In *Arabidopsis* and tobacco it has been demonstrated that homozygous and hemizygous plantlets can be selected in the T1 generation according to the intensity of GFP fluorescence (Fig. 2.1I; Niwa et al. 1999; Molinier et al. 2000). This distinction is easily observed with plants that have the GFP gene integrated at a single locus, while integration at multiple loci results in a more subtle differences in fluorescence intensities that are difficult to detect with the naked eye (Molinier et al. 2000). Simple observation of GFP fluorescence using an epifluorescence dissecting microscope permitted the selection of homozygous plants from hemizygous plants in the first and subsequent segregating generations (Molinier et al. 2000). Although

Table 2.2. Examples of GFP-based selection of segregating populations. *CM* Confocal microscope, *DM* dissecting microscope, *FM* fluorescent microscope

Plant	GFP type	Segregating material	GFP detection	References
Maize	58 mutations	Pollen	FM	Van der Geest and Petiolino (1998)
Arabidospsis	mtsGFP65T	Plantlets	CM	Niwa et al. (1999)
Tobacco	sGFPS65T	Seeds and plantlets	DM	Molinier et al. (2000)
Wheat	sGFPS65T	Embryos	FM	Jordan (2000)

selection by GFP fluorescence must be confirmed by proper molecular analysis, the reductions in greenhouse space and time spent are not negligible, compared to classification based on the genetic analysis of a large number of plants.

2.5 Conclusion

GFP-based selection of transgenic shoots has proven very useful in a large number of plant species. The properties of GFP enable this protein to be used as a very efficient in vivo selectable marker. The equipment required for its detection is not expensive and is available in most plant laboratories. Nevertheless, because many plant tissues naturally contain large amounts of fluorescent compounds, care must be taken to include the proper control experiments and to choose a well-adapted filter set in order to limit the occurrence of artifacts which may interfere with GFP fluorescence and render GFP useless as a selectable marker. Although GFP is already quite a powerful tool for the selection of transformed plants or tissues in certain cases, a better understanding of the factors limiting its use in e.g. direct regeneration systems is necessary before it can be considered as a universally applicable tool.

Acknowledgements. The authors would like to thank L. Valentine for critical reading of the manuscript, C. Himber for his technical assistance during the first transformation experiments performed using GFP, and Dr. R. Bronner for helpful discussions concerning the detection of GFP.

References

Ahlandsberg S, Sathish P, Sun C, Jansson C (1999) Green fluorescent protein as a reporter system in the transformation of barley cultivars. Physiol Plant 107:194–200

Blumenthal A, Kuznetzova L, Edelbaum O, Raskin V, Levy M, Sela I (1999) Measurement of green fluorescence protein in plants: quantification, correlation to expression, rapid screening and differential gene expression. Plant Sci 142:93–99

Chalfie M, Tu Y, Euskirchen G, Ward WW, Prasher DC (1994) Green fluorescent protein as a marker for gene expression. Science 263:802–805

Chiu W, Niwa Y, Zeng W, Hirano T, Kobayashi H, Sheen J (1996) Engineered GFP as a vital reporter in plants. Curr Biol 3:325–330

Davis SJ, Vierstra RD (1998) Soluble, highly fluorescent variants of green fluorescent protein (GFP) for use in higher plants. Plant Mol Biol 36:521–528

Elliott AR, Campbell JA, Dugdale B, Brettell RIS, Grof CPL (1999) Green-fluorescent protein facilitates rapid in vivo detection of genetically transformed plant cells. Plant Cell Rep 18:707–714

Ghorbel R, Juarez J, Navarro L, Pena L (1999) Green fluorescent protein as a screenable marker to increase the efficiency of generating transgenic woody fruit plants. Theor Appl Genet 99: 350–358

Haseloff J, Amos B (1995) GFP in plants. Trends Genet 11:328–329

Haseloff J, Siemering KR (1998) The uses of GFP in plants. In: Chalfie M, Kain SR (eds) Green fluorescent protein: properties, applications, and protocols. Wiley, Chichester, pp 191–220

Haseloff J, Siemering KR, Prasher DC, Hodge S (1997) Removal of a cryptic intron and subcellular localization of green fluorescent protein are required to mark transgenic *Arabidopsis* plants brightly. Proc Natl Acad Sci USA 94:2122–2127

Inouye S, Tsuji FI (1994) Aequorea green-fluorescent protein expression. Expression of the gene and fluorescence characteristics of the recombinant protein. FEBS Lett 341:277–280

Itaya A, Hickman H, Bao Y, Nelson R, Ding B (1997) Cell-to-cell trafficking of cucumber mosaic virus movement protein: green fluorescent protein fusion produced by biolistic gene bombardment in tobacco. Plant J 12:1223–1230

Jefferson RA, Ravanagh TA, Bevan MW (1987) GUS fusion: β-glucuronidase as a sensitive and versatile gene fusion marker in higher plants. EMBO J 6:3901–3907

Jordan MC (2000) Green fluorescent protein as a visual marker for wheat transformation. Plant Cell Rep 19:1069–1075

Kaeppler HF, Menon GK, Skadsen RW, Nuutila AM, Carlson AR (2000) Transgenic oat plants via visual selection of cells expressing green fluorescent protein. Plant Cell Rep 19:661–666

Leffel SM, Mabon SA, Stewart CN (1997) Applications of green fluorescent protein in plants. Biotechniques 23:912–918

Maximova SN, Dandekar AM, Guiltinan MJ (1998) Investigation of *Agrobacterium*-mediated transformation of apple using green fluorescent protein: high transient expression and low stable transformation suggest that factors other than T-DNA transfer are rate-limiting. Plant Mol Biol 37:549–559

Molinier J, Himber C, Hahne G (2000) Use of green fluorescent protein for detection of transformed shoots and homozygous offspring. Plant Cell Rep 19:219–223

Morise H, Shimomura O, Johnson FH, Winant J (1974) Intermolecular energy transfer in the bioluminescent system of *Aequorea*. Biochemistry 13:2656–2662

Nagatani N, Takumi S, Tomiyama M, Shimada T, Tamiya E (1997) Semi-real time imaging of the expression of a maize polyubiquitin promoter-GFP gene in transgenic rice. Plant Sci 124:49–56

Niwa Y, Hirano T, Yoshimoto K, Shimizu M, Kobayashi H (1999) Non-invasive quantitative detection and applications of non-toxic, S65T-type green fluorescent protein in living plants. Plant J 18:455–463

Ow DW, Wood KV, DeLuca M, de Wet JR, Helinski DR, Howell SH (1986) Transient and stable expression of the firefly luciferase gene in plant cell and transgenic plants. Science 234:856–859

Pang S-Z, DeBoer DL, Wan Y, Ye G, Layton JG, Neher MK, Armstrong CL, Fry JE, Hinchee MA, Fromm ME (1996) An improved green fluorescent protein gene as a vital marker in plants. Plant Physiol 112:893–900

Prasher DC (1995) Using GFP to see the light. Trends Genet 11:320–323

Sheen J, Hwang SB, Niwa Y, Kobayashi H, Galbraith DW (1995) Green fluorescent protein as a new vital marker in plant cells. Plant J 8:777–784

Vain P, Worland B, Kohli A, Snape J, Christou P (1998) The green fluorescent protein (GFP) as a vital screenable marker in rice transformation. Theor Appl Genet 96:164–169

Van der Geest AHM, Petiolino JF (1998) Expression of a modified green fluorescent protein gene in transgenic maize plants and progeny. Plant Cell Rep 17:760–764

Wang S, Hazelrigg T (1994) Implication for bcd mRNA localization from spatial distribution of exu protein in *Drosophila oogenesis*. Nature 369:400–403

Wintz H (1999) La GFP: structure, propriétés et applications. Regard Biochim 1:33–39

3 Luciferase Gene Expressed in Plants, Not in *Agrobacterium*

S.L. MANKIN

3.1 Introduction

The luciferase gene (*luc*) from the North American firefly *Photinus pyralis* has been used as a reporter gene in a variety of eukaryotic (Ow et al. 1986; Millar et al. 1992; Kost et al. 1995) and prokaryotic (Wood and DeLuca 1987) organisms. Luciferase assays are facilitated in plants by the lack of endogenous luciferase activity (Abeles 1986). The bioluminescent reaction catalyzed by luciferase can be visualized without sacrificing valuable tissue, and extraction of luciferase from plant tissues is simple and quick using commercially available assay kits. Detection of as few as 2,000 molecules of luciferase has been reported (Wood 1991). Although this review focuses upon the firefly luciferase, the techniques for visualizing bioluminescence and delivering substrate are applicable to other bioluminescence systems such as the *Renilla reniformis* luciferase system (Mayerhofer et al. 1995). Regardless of the system, bioluminescent activity is best detected using a combination of a photon-counting camera (Mankin et al. 1997) or cooled, slow-scan CCD camera (Kost et al. 1995) for tissue specificity and a luminometer for quantitative analysis (Millar et al. 1992; Mankin et al. 1997). Luciferase assay quantitation is facilitated by a broad linear range, which can extend over eight orders of magnitude (Wood 1991). Since the half-life of luciferase protein is quite short in the presence of luciferin (Nguyen et al. 1989; Thompson et al. 1991; Millar et al. 1992), reductions in mRNA levels are quickly converted into lower protein expression levels. Thus, luciferase provides a more dynamic indication of in vivo mRNA levels than the longer lived β-glucuronidase (GUS) or green fluorescent protein (GFP) reporter proteins (Ward and Bokman 1982; Narasimhulu et al. 1996); however, luciferase protein appears rather stable in the absence of luciferin and a GUS-like accumulated activity measurement can be made upon the first application of substrate (Millar et al. 1992).

3.2 Preventing Bacterial Expression

Plant promoters are often expressed in bacterial cells (Vancanneyt et al. 1990), and it takes weeks to eliminate *Agrobacterium* from plant tissues (Barghchi 1995; Suter-Crazzolara et al. 1995). Although *Agrobacterium* is a resilient

Molecular Methods of Plant Analysis, Vol. 22
Testing for Genetic Manipulation in Plants
Edited by J.F. Jackson, H.F. Linskens, and R.B. Inman
© Springer-Verlag Berlin Heidelberg 2002

pathogen that can last in soil for years, it is transmitted by wounding in contaminated soils (Agrios 1988) not by seed. During in planta transformation of *Arabidopsis thaliana*, *Agrobacterium* does not reappear when the seeds are surface-sterilized (Bechtold and Pelletier 1998; Clough and Bent 1998). Therefore, difficulties arise only when attempting to visualize luciferase expression during early stages of transformation, since any observed signal could derive from plant cells or from *Agrobacterium*; however, the presence of an intron in the luciferase transgene eliminates expression in bacterial cells while permitting normal expression in plants (Mankin et al. 1997). Although any intron should suffice, ideally it should contain several stop codons to prevent translation of unprocessed mRNA in bacterial cells, and the splicing machinery in plant nuclei should efficiently remove it. Plant introns are generally AT-rich, have an average length of 250 bp, and contain splice junctions similar to those in animal introns (Shapiro and Senapathy 1987; Goodall et al. 1991; McCullough et al. 1993). The luc^{INT} luciferase gene (Mankin et al. 1997) contains the second intron (PIV2) of the potato *ST-LS1* gene (Eckes et al. 1986) as modified by Vancanneyt et al. (1990). PIV2 is a typical plant intron with an 80% AT content, a length of 189 bp, typical splice junctions, and multiple stop codons in all translational reading frames. Vancanneyt et al. (1990) created the intron PIV2 from the second intron of ST-LS1 by altering the internal splice borders to match the consensus plant intron sequence and inserting it into the bacterial GUS gene. Transcripts of the resulting gus^{INT} gene are spliced effectively in *Arabidopsis* (Vancanneyt et al. 1990), tobacco (Vancanneyt et al. 1990; Rempel and Nelson 1995), and maize (Narasimhulu et al. 1996). The luc^{INT} gene has also been effectively expressed in tobacco (Mankin et al. 1997), *Arabidopsis,* maize and onion (SL Mankin, T Nguyen, WF Thompson), unpubl. data).

3.3 Imaging Luciferase Activity In Planta

Imaging luciferase expression in planta involves two main experimental factors: luciferin uptake and low-light imaging or bioluminescence quantification. Luciferin uptake has been achieved by a variety of methods, including spraying (Millar et al. 1992, 1995a, b), watering (Wood and DeLuca 1987), and petiole feeding (Barnes 1990). Bioluminescence of LUC-expressing plants has been visualized and/or quantified using X-ray film (Ow et al. 1986; Wood and DeLuca 1987), cooled, slow-scan CCD cameras (Kost et al. 1995), photon-counting cameras and luminometers (Mankin et al. 1997). Most low-light photodocumentation systems can be adapted for this purpose (Hooper et al. 1994); however, a light-tight imaging box is critical for keeping light noise to a minimum (Millar et al. 1992; Hooper et al. 1994). Determining the best methods for your needs requires some minor experimentation to establish the parameters for your equipment.

Ideally, a small pilot study should be initiated with LUC⁻ and LL⁺ plant material to establish the imaging procedure; however, if control material is unavailable, then a microtiter plate with a dilution series of recombinant luciferase standard is ideal for establishing the imaging parameters for your visualization system (Millar et al. 1992). Best results are obtained using an extractive assay buffer plus 1 mg/ml BSA to dilute the luciferase standard and a reaction buffer to supply substrates for the reaction since their combination is optimized to extend the luminescence plateau. Another easily overlooked parameter involves chlorophyll autofluorescence; therefore a dark incubation step of several minutes should be included prior to imaging luciferase activity in green tissues. When attempting to obtain quantitative images, results are more consistent with uniform tissues (e.g. 4-day-old seedlings) and treatments (e.g. 5-min transpiration of 1 mM D-luciferin followed by a 5-min dark incubation). Consistency of the treatment should be established by several replications with control material prior to attempting quantitative studies.

There are several proven methods for luciferin application. A 150 mM D-luciferin stock solution should be made, aliquoted, and stored in the dark at −70°C. Working solution (1–5 mM in 0.01% Triton X-100) may be stored for several hours in the dark at room temperature or overnight on ice. For tissues that lack well-developed cuticles, e.g. seedlings grown in sealed agar plates, spraying samples with a fine mist of working solution is very effective (Millar et al. 1992,1995a,b). Any mister or atomizer can be used, even a small, clean hairspray bottle. If sterile conditions need to be maintained, filter sterilize the working solution and sterilize the spray bottle by soaking in 70% EtOH prior to use. Samples should be sprayed evenly until they are completely coated with working solution; however do not drench them until they are submerged. Alternatively, luciferin can be delivered by watering plants with working solution (Wood and DeLuca 1987). This is most effective with smaller plants grown in media rather than soil. For large plants, the petiole feeding of individual leaves provides good results. In transgenic tobacco plants, essentially all the leaf interveinal areas were penetrated with luciferin after 20 min of petiole feeding; the luciferase luminescence remained stable between 20 and 60 min after application (Barnes 1990) and persisted for days in culture (data not shown).

Repeated application of luciferin is possible; however, the bioluminescence reaction of the firefly luciferase consumes ATP and O_2 as well as luciferin. In highly expressing young *Arabidopsis thaliana* seedlings repeatedly exposed to luciferin, the resulting strain can sometimes cause severe symptoms, including bleaching, arrested growth, or apical burns (data not shown) when grown in soil. (I do not recommend fertilizing immediately after treatment; in my hands this proved lethal.) Young seedlings (<2 true leaves) should be allowed to recover from treatment on media prior to transfer to soil to avoid loss.

A variety of methods have been developed to image bioluminescent samples. X-ray film is inexpensive and readily available; however, using film to visualize luciferase expression is relatively difficult due to the limited exposure ranges of film and the requirement of pressing samples flat onto the film. Low-light imaging using cooled, slow-scan CCD cameras (Kost et al. 1995) or VIM photon-counting cameras (Mankin et al. 1997) are ideal for analyzing luciferase expression in plant tissues. There are several complete systems available from commercial sources. A light-tight sample chamber is essential with these systems to reduce background light noise. Keep in mind that you might want to image whole, adult plants in the future, so imaging systems with minimal head space should be avoided. If you invest in a VIM or CCD camera system, take the time to learn how to use it properly, since many of the sensitive cameras can be damaged with improper use.

3.4 Measuring Luciferase Activity in Plant Extracts

Normally, a sharp peak of luminescence with a very rapid decline characterizes the reaction of luciferase, D-luciferin, ATP and O_2 (Wood 1991). This reaction can be extended to produce a "stable" luminescence plateau of 30- to 60-s duration by including coenzyme A in the reaction mix (Wood 1991; Ford et al. 1992). Luciferase extractive assay methods have been described in detail previously (Millar et al. 1992; Mankin et al. 1997), so only general principles will be covered here.

Unlike animal systems, firefly luciferase has a very short half-life in plant cell extracts (~2h at 0°C). Therefore, extra care should be taken to insure that all samples are treated similarly with timed steps and the extracts kept at 0–4°C. The short half-life appears to be dependent upon the source of tissue, the efficiency of extraction, and the extraction buffer (unpublished data), indicating that cellular proteinases are responsible. Electroporated protoplasts provide an excellent source tissue for studying luciferase protein decay in plant extracts because they offer consistent, complete extraction of soluble proteins. The shortened half-life excludes the use of microtiter-plate luminometers, which require 40–60 min to assay an entire 96-well plate. Many microtiter luminometers were designed specifically for GUS light assays and lack a broad enough linear range to handle a luciferase standard curve. Therefore, single- and dual-tube luminometers offer the best performance for firefly luciferase assays using plant extracts until a more stabilizing extraction buffer is determined. The optimal temperature of luciferase activity is 22°C (Promega 1993), but the instruments heat up over time – so temperature control for the sample chamber is an additional feature to consider. Regardless of the luminometer chosen, assays should be performed using automated injection of substrate to guarantee exact control of assay timing. A second injector is handy for addition of reaction-stopping buffers, especially in microtiter luminometers where light can bleed into neighboring wells.

When performing luciferase assays, results can be reported in relative light units (RLU); however, a standard and background should be checked periodically to insure that the instrument is functioning properly. Unfortunately, RLU cannot be used to correlate two separate experiments unless a standard curve covering the experimental range is available. Luciferase standard curves should span 7–10 orders of magnitude and should be run at least twice (start and finish) using recombinant luciferase (Roche or Promega). Standards are prepared in extraction buffer supplemented with 1 mg BSA/ml (Millar et al. 1992) and when prepared in this fashion are stable >10h at 0°C. Fresh dilutions should be made each day from a luciferase stock solution of 1 mg/ml. Stock solution should be aliquoted, stored at –70°C, and not exposed to repeated freeze-thaw cycles.

Acknowledgements. I would like to thank D. Arias for critically reading this manuscript. I would also like to thank William F. Thompson and Mary-Dell Chilton, whose rigorous training led to my better understanding science, in general, and luciferase assays, in particular.

References

Abeles F (1986) Plant chemiluminescence. Annu Rev Plant Physiol 37:49–72

Agrios G (1988) Plant Pathology, 3rd edn. Academic Press, New York

Barghchi M (1995) High-frequency and efficient *Agrobacterium*-mediated transformation of *Arabidopsis thaliana* ecotypes "C24" and "Landsberg erecta" using *Agrobacterium tumefaciens*. Methods Mol Biol 44:135–147

Barnes WM (1990) Variable patterns of expression of luciferase in transgenic tobacco leaves. Proc Natl Acad Sci USA 87:9183–9187

Bechtold N, Pelletier G (1998) *In planta Agrobacterium*-mediated transformation of adult *Arabidopsis thaliana* plants by vacuum infiltration. In: Martìnez-Zapater J, Salinas J (eds) *Arabidopsis* protocols: methods in molecular biology, vol 82. Humana Press, Totowa, NJ, pp 259–266

Clough S, Bent A (1998) Floral dip: a simplified method for *Agrobacterium*-mediated transformation of *Arabidopsis thaliana*. Plant J 16:735–743

Eckes P, Rosahl S, Schell J, Willmitzer •• (1986) Isolation and characterization of a light-inducible, organ-specific gene from potato *(Solarium tuberosum)* and analysis of its expression after tagging and transfer into tobacco and potato shoots. Mol Gen Genet 205:14–22

Ford SR, Hall MS, Leach FR (1992) Enhancement of firefly luciferase activity by cytidine nucleotides. Anal Biochem 204:283–291

Goodall GJ, Kiss T, Filipowicz W (1991) Nuclear RNA splicing and small nuclear RNAs and their genes in higher plants. Oxford Surv Plant Mol Cell Biol 7:255–296

Hooper C, Ansorge R, Rusbrooke J (1994) Low-light imaging technology in the life sciences. J Biolumin Chemilumin 9:113–122

Kost B, Schnorf M, Potrykus I, Neuhaus G (1995) Non-destructive detection of firefly luciferase (LUC) activity in single plant cells using a cooled, slow-scan CCD camera and an optimized assay. Plant J 8:155–166

Mankin SL, Alien GC, Thompson WF (1997) Introduction of a plant intron into the luciferase gene of *Photinus pyralis*. Plant Mol Biol Rep 15:186–196

Mayerhofer R, Langridge WHR, Cormier MJ, Szalay AA (1995) Expression of recombinant *Renilla* luciferase in transgenic plants results in high levels of light emission. Plant J 7:1031–1038

McCullough AJ, Lou H, Schuler MA (1993) Factors affecting authentic 5' splice site selection in plant nuclei. Mol Cell Biol 13:1323–1331

Millar AJ, Short SR, Hiratsuka K, Chua N-H, Kay SA (1992) Firefly luciferase as a reporter of regulated gene expression in higher plants. Plant Mol Biol Rep 10:324–337

Millar AJ, Carre IA, Strayer CA, Chua NH, Kay SA (1995a) Circadian clock mutants in *Arabidopsis* identified by luciferase imaging. Science 267:1161–1163

Millar AJ, Straume M, Chory J, Chua NH, Kay SA (1995b) The regulation ofcircadian period by phototransduction pathways in *Arabidopsis*. Science 267:1163–1166

Narasimhulu SB, Deng XB, Sarria R, Gelvin SB (1996) Early transcription of *Agrobacterium* T-DNA genes in tobacco and maize. Plant Cell 8:873–886

Nguyen VT, Morange M, Bensaude ·· (1989) Protein denaturation during heat shock and related stress. *Escherichia coli* β-galactosidase and *Photinus pyralis* luciferase inactivation in mouse cells. J Biol Chem 264:10487–10492

Ow DW, Wood KV, Deluca M, de Wet JR, Helinski DR, Howell SH (1986) Transient and stable expression of the firefly luciferase gene in plant cells and transgenic plants. Science 234:856–859

Promega (1993) luicferase assay system with reporter lysis buffer. Promega Tech Bull 161:1–8

Rempel HC, Nelson LM (1995) Analysis of conditions for *Agrobacterium*-mediated transformation of tobacco cells in suspension. Transgenic Res 4:199–207

Shapiro MB, Senapathy P (1987) RNA splice junctions of different classes of eukaryotes: sequence statistics and functional implications in gene expression. Nucleic Acids Res 15:7155–7174

Suter-Crazzolara C, Klemm M, Reiss B (1995) Reporter genes. Methods Cell Biol 50:425–438

Thompson JF, Hayes LS, Lloyd DB (1991) Modulation of firefly luciferase stability and impact on studies of gene regulation. Gene 103:171–177

Vancanneyt G, Schmidt R, O'Connor-Sanchez A, Rocha-Sosa M (1990) Construction of an intron-containing marker gene: splicing of the intron intransgenic plants and its use in monitoring early events in *Agrobacerium*-mediated plant transformation. Mol Gen Genet 220:245–250

Ward WW, Bokman SH (1982) Reversible denaturation *of Aequorea* green-fluorescent protein: physical separation and characterization of the renatured protein. Biochemistry 21:4535–4540

Wood KV (1991) Recent advances and prospects for use of beetle luciferases as genetic reporters. In: Stanley PE, Kricka J (eds) Bioluminescence and chemiluminescence: current status. Wiley, Chichester, pp 543–546

Wood KV, DeLuca M (1987) Photographic detection of luminescence in *Escherichia coli* containing the gene for firefly luciferase. Anal Biochem 161:501–507

4 Use of β-Glucuronidase To Show Dehydration and High-Salt Gene Expression

K. Nakashima and K. Yamaguchi-Shinozaki

4.1 Introduction

Plants respond to environmental stress, and the transduced signals cause expression of numerous genes associated with stress tolerance. A number of genes have been described that respond to water stress such as induced by drought and salinity in plants (Ingram and Bartels 1996; Bray 1997; Shinozaki and Yamaguchi-Shinozaki 1997, 1999, 2000; Hasegawa et al. 2000). More than 60 independent cDNAs for dehydration-inducible genes have been reported in *Arabidopsis* (Shinozaki and Yamaguchi-Shinozaki 1997, 1999, 2000). Functions of their gene products have been predicted from sequence homology with known proteins (Fig. 4.1). Genes induced during dehydration stress conditions are thought to function not only in protecting cells from dehydration by the production of important metabolic proteins (functional proteins) but also in the regulation of genes for signal transduction in the dehydration stress response (regulatory proteins). Northern analysis of dehydration-inducible genes revealed that there appear to be at least four independent signal-transduction pathways between the initial dehydration signal and gene expression (Fig. 4.2). Most of the dehydration-responsive genes are induced by the plant hormone abscisic acid (ABA), but others are not.

To understand the signal transduction pathways from the perception of the water-stress signal to gene expression, it is important to analyze the expression of the genes involved. The putative promoters of these genes can be fused to the β-glucuronidase reporter gene *(GUS)* to analyze the regulatory elements involved in stress-responsive transcription in transgenic plants (Fig. 4.3). By using the promoter-*GUS* system, areas of the promoter region that are important in stress-inducible gene expression can be examined. Moreover, the key *cis*-acting elements that are important in the response to water stress at the nucleotide level can be identified. For example, our group has succeeded in identification of a *cis*-acting element that is important in the response to stress. Promoter analysis using the *GUS* reporter system showed that a 9-bp conserved sequence, 5'-TACCGACAT-3', termed the dehydration-responsive element (DRE), is essential for the induction of the *rd29A* gene (also known as *cor78* and *lti78*) of *Arabidopsis* not only by osmotic stress, such as occurs in drought and high salinity, but also by low-temperature stress (Fig. 4.2). We also identified *cis*-acting elements in two other stress-inducible genes of *Arabidopsis* by utilizing a similar approach: the MYB (MYBRS) and MYC (MYCRS)

Molecular Methods of Plant Analysis, Vol. 22
Testing for Genetic Manipulation in Plants
Edited by J.F. Jackson, H.F. Linskens, and R.B. Inman
© Springer-Verlag Berlin Heidelberg 2002

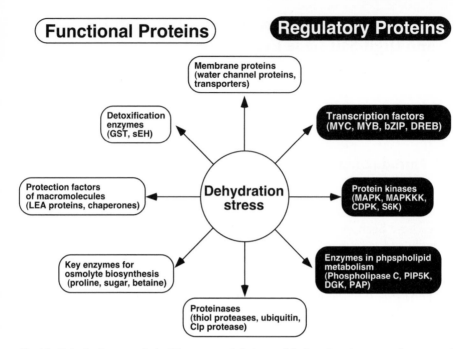

Fig. 4.1. Dehydration-tress-inducible genes and their possible functions in stress tolerance and response. Gene products are classified into two groups. The first group includes proteins that probably function in stress tolerance (functional proteins), and the second contains protein factors involved in regulation of signal transduction and gene expression that probably function in stress response (regulatory proteins)

recognition sequences in the *rd22* gene (Abe et al. 1997), and ABRE in *rd29B* (or *lti65*) gene (Uno et al. 2000). Promoter analysis of other dehydration-inducible genes, *erd1* (Nakashima et al. 1997), the P5C synthase gene (*P5CS*) (Yoshiba et al. 1999), and the DRE-binding protein 2 gene (*DREB2*) (Nakashima et al. 2000) is currently being conducted using the promoter-*GUS* system. Furthermore, the promoter-*GUS* system has also been used to analyze expression of the dehydration-repressive and rehydration-inducible proline dehydrogenase (*ProDH*) gene (Nakashima et al. 1998).

In this chapter, application of the promoter-*GUS* system in studies of gene expression in response to dehydration and salt stress in plants will be demonstrated. For more detailed information regarding GUS assay methods , we recommend several review articles (Jefferson et al. 1987; Jefferson and Wilson 1991; Hull and Devic 1995), and the book, *GUS protocols: using the GUS gene as a reporter of gene expression* (Gallagher 1992).

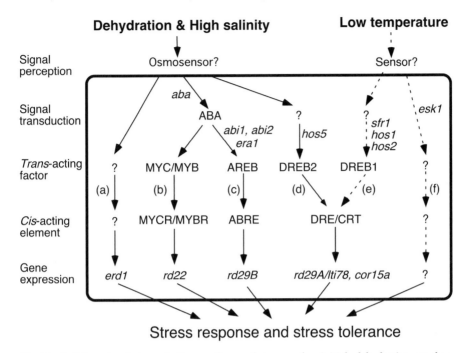

Fig. 4.2. Cellular signal transduction pathways between the initial dehydration or low-temperature stresses and gene expression in *Arabidopsis* (adapted from Shinozaki and Yamaguchi-Shinozaki 2000). There are at least six signal transduction pathways: two (*b, c*) are abscisic acid (ABA)-dependent and four (*a, d, e, f*) are ABA-independent. Stress-inducible genes *rd29A/cor78/lti78, rd29B/lti65, rd22,* and *erd1* have been used to analyze regulation of the signaling process. *abi1, abi2,* and *era1* are involved in ABA signaling; *hos5* functions in DREB2-related dehydration signaling; and *sfr6, hos1,* and *hos2* functions in DREB1/CBF-related cold signaling. *esk1* is involved in the response to cold via a DRE-independent process. *Thin arrows* represent the signaling pathways that are involved in dehydration-responsive *erd1* gene expression. *Broken arrows* represent the signaling pathways that are involved in low-temperature stress responses

4.2 What Is GUS?

Richard Jefferson et al. (1987) demonstrated the application of the *GUS* gene (*uidA* or *gusA*) of *Escherichia coli* as a reporter gene in transgenic plants. Since then, this gene has become one of the most widely used reporter genes in plant molecular biology.

The GUS enzyme hydrolyzes 5-bromo-4-chloro-3-indolyl β-D-glucuronic acid (X-Gluc) to a water-soluble indoxyl intermediate that is further dimerized into a dichloro-dichloro-indigo blue precipitate by an oxidation reaction (Fig. 4.4).

The GUS system has many advantages, as described by Jefferson and Wilson (1991). These include:

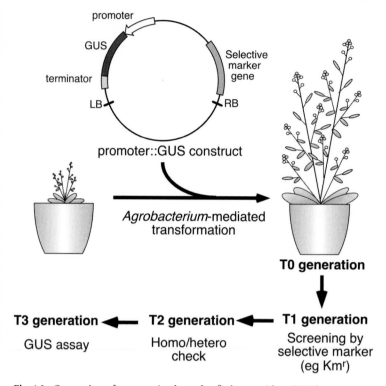

Fig. 4.3. Generation of transgenic plants for β-glucuronidase (GUS) assay

1. The GUS enzyme catalyzes the hydrolysis of a wide variety of β-glucuronides. Substrates are commercially available that allow GUS activity to be measured easily and quantitatively by spectrophotometric and fluorometric methods.
2. Endogenous GUS activity is absent in most higher plants, allowing the detection of low levels of GUS activity in tissues of transgenic plants. However, several reports have described endogenous GUS activity in nontransformed plants, notably in the male gametophyte (Iturriaga et al. 1989; Jefferson and Wilson 1991). A sample of tissue from nontransgenic plants should always be assayed as a control for background activity.
3. The tissue-specific localization of GUS activity can be visualized in a histochemical assay using the indigogenic substrate X-Gluc (Fig. 4.4).
4. GUS is a stable enzyme that can be assayed in simple buffers over a range of pH values (5.0–9.0).
5. GUS can tolerate large N-terminal and C-terminal fusions. This property has been utilized to study the function of signal peptides by generating GUS fusion proteins targeted to cellular organelles, such as the nucleus.
6. GUS mRNA is sufficiently stable to be detected by Northern analysis.

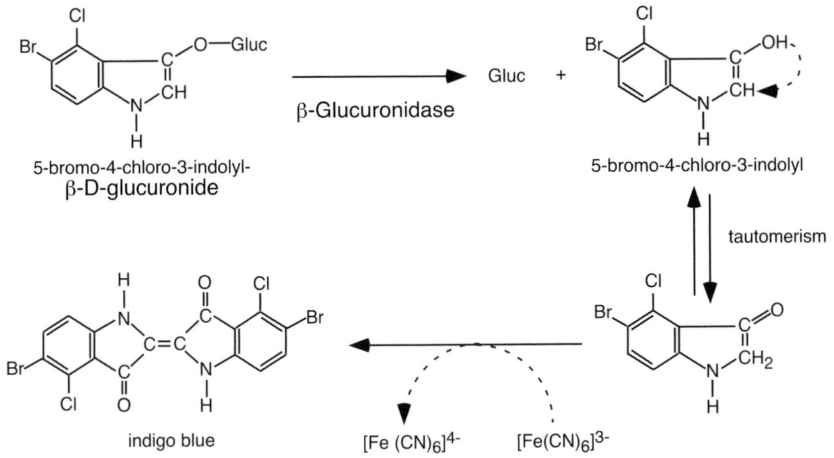

Fig. 4.4. Chemistry of X-Gluc reaction. Hydrolyzation of 5-bromo-4-chloro-3-indolyl β-D-glucuronic acid (X-Gluc) by the β-glucuronidase enzymes results in a reactive indoxyl mole cules (adapted from De Block and Van Lijsebettens 1998). Two indoxyl molecules are oxidized to indigo blue; ferri (III) cyanide enhances the dimerization

The *GUS* gene is frequently used as a reporter gene for promoter analysis, in both transient assays and in stably transgenic plants. It has been used to identify *cis*-acting elements involved in the regulation of gene expression at many levels, such as tissue-specific and developmental regulation, hormonal regulation, and stress-responsive regulation. Figure 4.5 shows a typical promoter-*GUS* fusion gene construct in T-DNA (Fig. 4.5A) and examples of a GUS assay: fluorometric analysis (Fig. 4.5B), Northern analysis (Fig. 4.5C), and histochemical analysis (Fig. 4.5D). Recently, other reporter genes have become available, such as luciferase (LUC) from fireflies and green fluorescent protein (GFP) from jellyfish (Millar et al. 1992). Although the LUC system is more sensitive than the GUS system, the *LUC* gene product is unstable. Therefore, the timing of the assay is difficult to select. On the other hand, since LUC can be detected in intact plants, the promoter-*LUC* system can be used for the screening of mutants that show altered signal transduction pathways from stress to gene expression (Ishitani et al. 1997). The GFP system can be used to detect the localization of gene expression at the microscopic level (Sheen et al. 1995). These reporter gene systems are discussed elsewhere in this volume.

4.3 Transgenic Plants Carrying Promoter-*GUS* Constructs

The first step is to search for stress-inducible genes, analyze the expression pattern by Northern blot or RT-PCR, and then isolate the promoter region of the gene in interest. These steps are summarized in a flow diagram in Fig. 4.6.

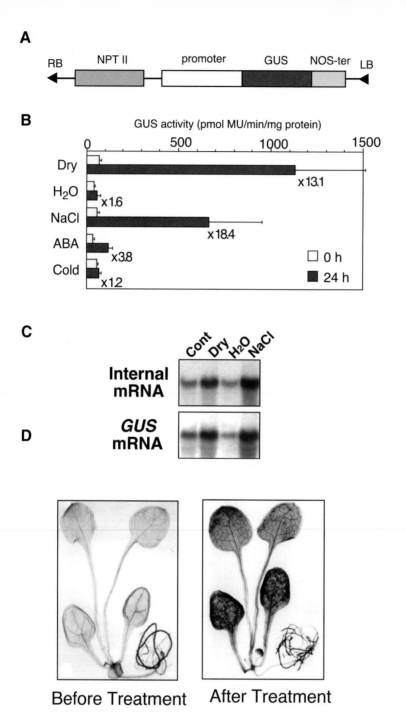

Fig. 4.5A–D. Examples of the GUS assay. **A** Typical construct of promoter-*GUS* fusion gene in T-DNA. **B** A GUS fluorometric assay using transgenic plants containing the promoter-*GUS* construct. **C** Northern analysis of transgenic plants containing the promoter-*GUS* construct to detect internal mRNA and GUS mRNA. **D** Histochemical analysis of transgenic plants containing the promoter-*GUS* construct by GUS staining

Searching for stress-inducible genes

Differential screening of cDNA
Differential display
Homology-based PCR cloning
Microarray or macroarray

Northern blot or RT-PCR analysis

Check the stress inducibility
Control condition (without stress)
Dehydration stress (1,2,5,10,24 h)
High salinity stress (250 mM NaCl; 1,2,5,10,24 h)
ABA treatment (100 μM ABA; 1,2,5,10,24 h)
Cold stress (4°C or 0°C; 1,2,5,10,24 h)
 and other stresses such as heat stress (42°C)

Check the Developmental Expression
Root
Immature younger leaves
Mature older leaves
Stem
Flower
Fruit (Silique)
Matured seed
 and in situ hybridization when you want to see in detail

Isolation of promoter region

When the genomic data is available
PCR-based cloning of the promoter region (usually ca. 2-0.5 kb)

When the genomic data is not available
Cloning of genomic DNA from genomic library

Construction of promoter-GUS

Design and make the promoter constructs (pSK etc.)
Sequence check
Ligation of the promoter and the GUS-containing binary vectors (pBI etc.)
Introducing the promoter-GUS constructs in *Agrobacterium*

Introduction of promoter-GUS constructs into plants

Arabidopsis: Infiltration (Vacuum or flower dipping or spray)
Tobacco and rice: co-culture

GUS assay using the transgenic plants containing the promoter-GUS constructs

To see the stress inducibility of the promoter
Fluorometric assay
Northern analysis using the GUS gene as probe

To see the tissue specificity of the promoter
Histochemistry

Fig. 4.6. Promoter analysis using the promoter-*GUS* system

In this section, the steps to make the promoter-*GUS* construct and to generate transgenic plants, especially *Arabidopsis*, carrying the promoter-*GUS* construct, are described.

4.3.1 Construction of Promoter-*GUS* Fusion Genes

A number of vectors designed for the construction of *GUS* gene fusions are described in Jefferson (1988), and can be purchased from Clontech (Palo Alto, CA). These include the pBI101 series, in which the *GUS* gene coding sequence is cloned into the binary vector pBin19, and flanked by the *lacZ* polylinker sequence at the 5′ end and the 3′ termination sequence of the nopaline synthase (*nos*) gene of *Agrobacterium tumefaciens* at the 3′ end. This allows DNA sequences to be inserted upstream of the *GUS* gene, generating gene fusions that can be transferred directly into plants by *Agrobacterium*-mediated transformation.

Two types of *GUS* gene fusions can be constructed: transcriptional or translational gene fusions. In transcriptional gene fusions, the site of the fusion is upstream of the ATG initiation codon of the *GUS* gene. The transcriptional gene fusions are usually generated for promoter analysis. In translational gene fusions, the site of the fusion is within the translated DNA sequence. In this case, it is essential that the correct reading frame is conserved across the junction of the two coding regions. For the construction of translational fusions in the N-terminal region, Clontech offers three kinds of pBI101 vectors, pBI101.1, pBI101.2, and pBI101.3, each of which provides a different reading frame relative to the cloning site.

DNA fragments to be cloned upstream of the *GUS* gene can be generated by several strategies. First is utilization of the existing restriction sites, second is utilization of the deleted sequences by exonuclease III digestion, and third is PCR. In the case of PCR, a DNA polymerase with high fidelity (low error rate), such as KOD DNA polymerase (Toyobo, Osaka, Japan) or *Pfu* Turbo DNA polymerase (Stratagene, CA) is recommended to generate the DNA fragment. When PCR is used to generate the DNA fragment, it is important to check the sequence of the amplified fragment. Once the fragments are cloned into the GUS vectors, it is also necessary to check the sequence of the junction site. In our laboratory, when the promoter DNA fragment contains suitable restriction sites for cloning, the DNA fragment is digested at these restriction sites and cloned into the pBI101 vector. However, in many cases promoter DNA fragments do not have the suitable restriction sites. In these cases, we use PCR to generate DNA fragments by making a "5′ primer" that contains *Hind* III or *Sal* I sites upstream of the 5′ side of the promoter and a "3′ primer" that contains a *Bam*HI site downstream of the 3′ side of the promoter. Addition of three to five random nucleotides at the 5′ side of the primers is necessary to digest the terminal restriction sites of the PCR fragments completely. When the amplification product is generated by *Pfu* DNA polymerase, promoter-DNA

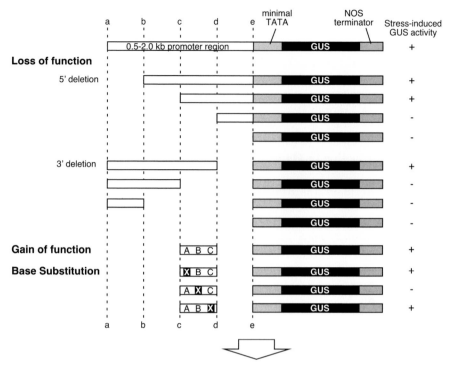

Fig. 4.7. Loss-of-function analysis, gain-of-function analysis, and base substitution to identify the stress-inducible *cis*-acting element

fragment can be subsequently cloned into pBluescript II SK+ (pSK)-type plasmids using pPCR-Script kit (Stratagene, CA). The sequence of the amplified fragment should be checked using an M13 primer and a reverse primer. Then the cloned promoter fragment cut by restriction enzymes is re-cloned into the upstream cloning site of the pBI101 vector.

To identify the stress-inducible *cis*-acting element, loss-of-function analysis, gain-of-function analysis, and base-substitution experiments should be conducted. The constructs for these experiments are shown in Fig. 4.7 (for further information on this kind of experiment, see Yamaguchi-Shinozaki and Shinozaki 1994; Abe et al. 1997; Liu et al. 1998; Uno et al. 2000).

4.3.2 Introduction of Promoter-*GUS* Constructs into *Agrobacterium*

Plasmids pBI101 containing the promoter-*GUS* fusion constructs were transferred from *E. coli* DH5α into *Agrobacterium tumefaciens* by electroporation using Bio-Rad Gene Pulser or via triparental mating with an *E. coli* strain that contained the mobilization plasmids pRK2013. *Agrobacterium* containing the construct should always be maintained in a glycerol stock at −80°C, since *Agrobacterium* cannot survive on LB medium at 4°C longer than 2–3 weeks. For tandem-repeat promoter constructs, we recommend that a glycerol stock be made as soon as possible. It has been reported that homologous recombination sometimes occurs in *Agrobacterium* cells at 4°C, and copy number decreases after 1 month (Weigel et al. 2000).

4.3.3 Transformation of Plants with *Agrobacterium*

4.3.3.1 Transformation of *Arabidopsis* Plants

In this section we will introduce the protocols for three kinds of infiltration methods for transforming *Arabidopsis* plants with *Agrobacterium* containing the constructed promoter-*GUS* fusion gene; vacuum infiltration (modified from Bechtold and Pelletie 1998), floral dip (modified from Clough and Bent 1998), and spray methods (modified from Weigel et al. 2000). The infiltration method is based on the assumption that the gene-transfer stage is at the end of gametogenesis or at the zygote stage. This assumption was based on the observation that transformants are hemizygous for T-DNA insertions (Bechtold and Pelletie 1998). Regardless of the infiltration method that is chosen, the most important thing is to use *healthy* plants.

First, the plants must be prepared. Usually, we use *Arabidopsis thaliana* (L.) Heyn., ecotype Columbia (Col-0). Ecotypes Wassilevskija (Ws-0), Nossen (No-0), and Landsberg *erecta* (L*er*) may also be used with good efficiency (Becktold and Pelletie 1998). Dip 20 mg (1,000 seeds) of wild-type seeds in 10 ml of distilled water (dH$_2$O) in a 15-ml tube, Keep the tube in the cold room (4°C) for 2–3 days. Replace the dH$_2$O with 0.1% agarose. Spread 12–16 seeds on a surface of soil in a 4-inch pots using a Pasteur pipette or a 1-ml scale Pipetman (Gilson). For the vacuum or dipping methods, the soil should be covered with nets, but for the spray methods, there is no need to cover the soil. Two to four plants can be vacuum-infiltrated with 250 ml of *Agrobacterium* in the infiltration medium. Four to six pots can be dipped into 250 ml of *Agrobacterium* in the infiltration medium. Eighteen to twenty-four pots can be sprayed with 250 ml of *Agrobacterium* in infiltration medium. Place the pots in a flat covered with a transparent dome or Saran Wrap. Transfer the pots to a greenhouse (long day, low humidity) or a growth chamber (16 h light/8 h dark, 22°C). After 1 week, remove the cover. In 4 weeks the plants will be old enough to be clipped (primary inflorescence should be about 10 cm tall). Clip the plants at the base of the stem 3–6 days before the infiltration. On the day of infiltration, the plants should have many buds and flowers and almost no siliques. If there are open flowers and siliques, cut and remove them. Water plants thoroughly a couple of days before the infiltration of *Agrobacterium*. Do not overwater immediately before infiltration.

On the first day of the infiltration, prepare 500 ml of YEP liquid culture. Inoculate 5 ml of liquid LB plus antibiotic with a single fresh colony or a −80°C glycerol stock of the *Agrobacterium* containing the promoter-*GUS* construct. The typical *Agrobacterium* strain used is C58 (rifampicillin, 100 μg/ml) or GV3101 pMP90 (Gentar, 25 μg/ml). A typical vector such as pBI101 series is usually Kan^r (25–60 μg/ml). On the second day, dilute the overnight liquid culture 1:1,000 in a large flask (2-l) containing 500 ml of the liquid medium (YEP plus antibiotic). Since *Agrobacterium* needs good aeration for growth, a smaller flask is not recommended. Grow the culture overnight at 28°C with shaking (160–200 rpm). On the third day, prepare the infiltration medium. To 1 l of distilled water, add 0.5× MS salts, 50 g sucrose (5%). Stir well. Adjust pH to 5.0 with KOH. Add 1 ml of 1,000× Gamborg's vitamin stock, and sterilize in an autoclave. Before use, add 2 μl of 1 mg benzylaminopurine (BA)/ml in DMSO (10 mg/l final concentration, store at −20°C, use within 1 month), and 62.5–100 μl of Silwet L-77 (final concentration should be 0.025–0.04%) in 250 ml of infiltration medium. Retrieve the liquid culture from the shaker and read the OD$_{600}$; it should be higher than 1.5. Pour the culture into centrifuge containers and centrifuge at 5,000 rpm for 10 min at 10–20°C (not 4°C!). Resuspend the cells in infiltration medium; this should be done gently, without creating too much foam. Adjust the OD$_{600}$ of the cells in the infiltration medium to about 1.0. We usually suspend the cells from a 500-ml culture in 250 ml of infiltration medium.

4.3.3.1.1 Vacuum Infiltration

Fill a 200-ml beaker with infiltration medium. Dip the aerial parts of the plants in the *Agrobacterium* suspension. The same suspension can be used several times. Place the beaker with plants and suspension in a 10-l vacuum chamber and apply 60–70 cmHg of vacuum pressure for 10 min. Gently release the vacuum and remove the beaker.

4.3.3.1.2 Floral Dip

Pour the infiltration medium into large, tall Petri dishes or a plastic container. Dip the aerial parts of the plants in the *Agrobacterium* suspension for 10–20 s. Swirl the plant around gently. You may see a film of liquid coating the plant. Remove the pot and lay it on its side inside a flat. Although the transformation efficiency of the flower dipping method is lower than the vacuum method, it is easier and more stable (the vacuum process sometimes affects the plants and *Agrobacterium*).

4.3.3.1.3 Spray

Spray the infiltration medium containing *Agrobacterium* cells onto the *Arabidopsis* plants. Although the transformation efficiency of the spray methods is lower than the flower dipping method, it is very easy.

After dipping and spraying, cover the flats with a transparent plastic dome or wrap. Keep the plants away from direct light to reduce damage to the plants by overheating under the dome. The following day, uncover the plants and return them to an upright position. Do not water plants for at least 2 days. For both the dipping and spraying methods, the whole process should be repeated 1 week after the first transformation. Following treatment, the leaves dry rapidly but the floral stems become erect and continue to flower. Water the plants moderately until maturity (approx. 4 weeks) and let the plant dry progressively. Harvest the seeds (T1 seeds) from the plants in bulk after the silique is completely dry.

The transformed seeds (T1) must be germinated on kanamycin selection plates if Kan^r is used as a selection marker. Prepare plates for kanamycin selection. One liter of kanamycin selection medium contains 0.5× MS salts, 0.8% bacto-agar. After autoclaving, add 1× Gamborg's vitamins, and kanamycin sulfate to a final concentration of 35–60 μg/ml. Pour the media onto the plates (diameter 15 cm). Then sterilize the seeds for kanamycin selection. The density of seeds on one plate is approximately 2,000–3,000 seeds (40–60 mg). Place seeds in 15-ml plastic tubes and treat for 1 min in 70% ethanol, 5–10 min in 50% bleach, 0.05% Tween 20 or Triton X-100; rinse them three times in sterile dH$_2$O under the hood using autoclaved pipettes. Suspend the sterilized seeds in a sterile 0.1% agarose solution (5 ml/plate) and spread them onto the plate. Dry the plates in the hood until seeds on the plate no longer move when the plate is tipped. Tape the plates with tissue culture tape and vernalize plates for 2 days at 4°C. Move the plates to the growth chamber (16 to 24 h light). After about 2 weeks, the transgenic plants that survive on the plates containing the selection marker can be selected. Carefully transfer the transgenic plants to soil and keep them in high moisture conditions for 3–5 days (cover with a dome or Saran Wrap). Then culture the plants in the greenhouse or the growth chamber to obtain T2 seeds. The T2 plants for can be used for GUS assay more than 20 plants are analyzed; however, bear in mind that the T2 plants may contain homozygous and heterozygous transgenes, and the insertion of T-DNA may be at a single locus or at more than two loci. We recommend using homozygous single-locus T3 or T4 transgenic plants for GUS assay (Fig. 4.3).

Segregation analysis can be carried out using the T2 generation. Sterilize the surface of the seeds in 70% ethanol and a diluted bleach solution, and plant out on sterile growth media containing the appropriate selection agent. The T2 generation is then scored for the number of resistant seedlings compared to sensitive seedlings for different numbers of insertional events. If the T-DNA insertion is at one locus, the ratio of resistant to sensitive should be 3:1. If the T-DNA insertion is at two loci, the ratio of resistant to sensitive should be 15:1. And if the T-DNA insertion is at more than two loci, the ratio of resistant to sensitive should be >15:1. However, if the T-DNA insertion is at one locus, causing an embryo-lethal, the ratio of resistant to sensitive should be 2:1 (75% germination). When the insertion causes lethal gametophilic mutation, all seeds will be sensitive (25% germination). Then the plants with the insertion at one locus should be transplanted in soil and the T3 seeds harvested. By checking the segregation ratio of T3 seeds on the media containing the selection agent, homozygous transgenic plants can be selected. If all of the T3 seeds of one line are able to grow in medium containing the selection marker, this line should be homozygous.

4.4 Fluorometric Assay

4.4.1 Introduction

We assay GUS activity in tissue extracts by fluorometric quantitation of 4-methylumbeliferone produced from the glucuronide precursor using the method described by Jefferson et al. (1986). This protocol can be used for assaying the expression of GUS in cells from stably transformed plants. Plant material can be frozen prior to assay at −80°C without any loss in activity. However, do not freeze samples in the GUS buffer, as this will result in significant loss of activity (Topping and Lindsey 1997). Here, we will describe the method for transgenic *Arabidopsis* plants containing promoter-*GUS* constructs.

4.4.2 Stress Conditions

4.4.2.1 Plant Preparation

Arabidopsis plants are grown on germination medium (GM) agar plates (Valvekens et al. 1988) for 3–4 weeks (Yamaguchi-Shinozaki and Shinozaki 1994).

4.4.2.2 Dehydration

Arabidopsis rosette plants are harvested from GM agar plates and then dehydrated in plastic culture dishes without covers at 22°C in 60–70% humidity under dim light.

4.4.2.3 High Salinity

Arabidopsis rosette plants are transferred from GM agar plates to hydroponic growth in 250mM NaCl or dH$_2$O (control) in plastic culture dishes at 22°C under dim light.

4.4.2.4 ABA Treatment

Arabidopsis rosette plants are transferred from GM agar plates to hydroponic growth in 10–100 µM ABA or dH$_2$O (control) in plastic culture dishes at 22°C under dim light. Otherwise, plants should be sprayed with 10–100 µM ABA or dH$_2$O, then left at 22°C under dim light

4.4.2.5 Other Treatments

To test whether osmotica induces gene expression, 20–30% polyethylene glycol (PEG) and 0.4–0.6 M mannitol are used. In these cases, *Arabidopsis* rosette plants are transferred from GM agar plates to hydroponic growth in 20–30% PEG or 0.4–0.6 M mannitol in plastic culture dishes at 22°C under dim light. Transfer the plants from agar to dH$_2$O as a control.

For cold treatment, incubate the plates containing *Arabidopsis* rosette plants in a cold-room (4°C) or an incubator (0°C). For heat treatment, incubate the plates containing *Arabidopsis* rosette plants in an incubator at 37 or 42°C.

4.4.3 Protein Assay

First prepare a calibration curve using bovine serum albumin (BSA). Prepare a solution of 1 mg BSA/ml and GUS extraction buffer (50 mM sodium phosphate, pH 7.0, 1 mM EDTA, 0.1% Triton X-100, 10 mM 2-mercaptoethanol; 2-mercaptoethanol should be added to GUS extraction buffer just before use). Prepare Bio-Rad protein assay reagent. To make 25 ml protein assay solution (for about 25 assays), mix 5 ml of the original Bio-Rad protein assay reagent, 20 ml distilled water, and 50 µl 1 M Tris-HCl, pH 8.0.

Aliquot 1, 2, 5, and 10 µl of the BSA solution in cuvette. Add 5 µl of GUS extraction buffer (a volume of the GUS extraction buffer equal to that of the plant extract to be assayed should be added to the BSA standards, as some constituents of the GUS extraction buffer might absorb light at 595 nm). Add 1 ml of the diluted protein assay solution and mix thoroughly. Leave at room temperature (20–25°C) for 5 min. Measure the absorbance at 595 nm and plot absorbance against amount of protein (µg).

4.4.4 Sample Preparation

Put a leaf disc or an unbolted transgenic plant into a 1.5-ml Eppendorf centrifuge tube and grind thoroughly in 150 µl of GUS extraction buffer using a homogenizer or motorized micropestle. Remember to assay nontransformed plants or plants transformed with the vector only. Put the tubes on ice. Centrifuge the homogenate in an Eppendorf tube at maximum speed (~15,000 rpm) at 4°C for 5 min, then place on ice.

Put 5 µl of the supernatant in a 1.5-ml disposable microcuvette (Kartell, Italy, or equivalent). Add 1 ml of the diluted protein assay solution, and mix thoroughly. Leave at room temperature for 5 min. Measure the absorbance at 595 nm and determine the concentration of protein in the supernatant using the calibration curve.

4.4.5 Fluorometric Assay

Prepare 4-MU standards (GUS extraction buffer containing 100 pmol, 1 nmol, 10 nmol, and 100 nmol of 4-MU), 1 mM MUG solution (1 mg of MUG in 2.5 ml of GUS extraction buffer; make up freshly), and 0.2 M Na_2CO_3.

Aliquot 5 µl of the supernatant to a 10-ml disposable test tube. For the calibration, add 5 µl of GUS extraction buffer containing 0 nmol, 100 pmol, 1 nmol, 10 nmol, and 100 nmol of 4-MU. Add 250 µl of 1 mM MUG to each tube and mix well. Incubate the test tubes at 37°C for 1 h. Add 2.5 ml of 0.2 M Na_2CO_3. During the incubation, switch on the fluorometer (Hitachi Fluorescence Spectrophotometer F-2000), as it requires at least 30 min to warm up. Measure the fluorescence units of the samples (excitation: 365 nm, emission: 455 nm). Using the 4-MU calibration curve, convert the rate of GUS activity from fluorescence units per min into pmol 4-MU per min.

It is important to ensure that the plant tissue used for the GUS assay is free of *Agrobacterium*, which will give false-positive results. The presence of *Agrobacterium* can be determined by grinding up the tissue sample and incubating it in an LB medium overnight at 28°C.

4.5 Histochemistry

4.5.1 Introduction

In stably transformed plants, the expression pattern of GUS gene fusions can be analyzed very precisely, both temporally and spatially by a histochemical assay (Topping and Lindsey 1997). The GUS activity can be localized accurately in vivo using X-Gluc as a substrate (Fig. 4.4).

The advantages in using GUS for this type of experiment include:

1. A low background GUS activity in the majority of plant tissues (see Wilkinson et al. 1994).
2. The product is non-diffusible and accumulates within the cells where the gene is expressed.
3. The substrate is readily taken up into the plant cells by simple diffusion or following vacuum infiltration of tissues.

There are, however, some disadvantages (Topping and Lindsey 1997).

1. The assay is not readily suitable for use on living tissues.
2. The GUS enzyme and product are very stable and can therefore give an unrealistic picture of the steady-state abundance of the GUS protein. Thus, the assay is less accurate in determining the time that gene expression is switched off.
3. In tissues in which there is a high level of GUS expression, some leakage of the colorless reaction intermediate has been shown to occur; however, this can be minimized by the addition of either, or both, of the oxidative agents potassium ferri- and ferrocyanide to a final concentration of up to 5 mM in the substrate solution.
4. The results are not quantitative.

4.5.2 Histochemistry

Prepare FAA (5% formaldehyde, 5% acetic acid, and 37.5% ethanol), 1 mM X-Gluc buffer (50 mM sodium phosphate buffer, pH 7.0, 5–7% methanol, 0.5 mg X-gluc/ml) and ethanol series (50, 70, 90, and 100%).

Fresh sections (80–100 μm) of plant tissue can be cut with a Microslicer (DTK-1000; Dosaka EM, Kyoto, Japan) from tissue that has been embedded in 5% agar.

Put three to five transgenic, unbolted plants or the fresh sections in dH_2O and apply a vacuum for 2–3 min. Remove the solution and add 2 ml of 1 mM X-Gluc buffer to the plants. Incubate the plants in at 37°C for 0.5 to 24 h. Fix the plant tissues in FAA, then remove the chlorophyll with 50% ethanol for 5 min, 70% ethanol for 24 h, 90% ethanol for 5 min, and 100% ethanol for 5 min. Keep the plants in dH_2O and take pictures using a stereo microscope.

4.6 Northern Analysis of GUS

4.6.1 Introduction

If GUS enzyme activity cannot be detected, it may be possible to detect the accumulation of GUS mRNA by Northern analysis. The RNA is separated according to size by electrophoresis through a denaturing agarose gel and subsequently transferred to a nylon membrane (Sambrook and Russell 2001). A radiolabeled DNA probe of the GUS gene is then hybridized to the accumulated GUS mRNA. The level of hybridized probe can be visualized by autoradiography. We recommend that the mRNA of the gene of interest (the gene that was used for promoter analysis) be detected on the same blot by using a mixed probe. In RNA work, including RNA preparation and Northern blotting, RNase contamination of the solutions and apparatuses used for Northern analysis must be prevented. DEPC (diethyl pyrocarbonate) treatment of solutions, including H_2O, is effective for inactivating RNase. Caution: DEPC is a suspected carcinogen and should be handled with care. Wherever possible, the solutions should be treated with 0.1% DEPC for at least 12h at 37°C and then autoclaved for 20 min (Sambrook et al. 1989). DEPC reacts rapidly with amines and cannot be used to treat solutions containing buffers such as Tris.

4.6.2 RNA Extraction

There are several kinds of RNA extraction methods. Here we describe the ATA method, which is suitable for Northern analysis. ATA, aurintricarboxylic acid, is a strong inhibitor of the interaction between nucleic acid and protein and is an RNase inhibitor. The ATA method is relatively inexpensive and the degradation of RNA is low. Therefore, RNA can be extracted from 30–50 samples at a time. However, ATA inhibits the activity of reverse transcriptase, so the RNA extracted by this method cannot be used for RT-PCR and cDNA synthesis. Trisol (Gibco BRL) is an excellent, albeit expensive, reagent for extracting RNA. However, Trisol-extracted RNA can be used for RT-PCR as well as Northern analysis. To use TRISOL, see the protocols provided by the manufacturer.

Before and after stress treatment, freeze 0.5–2g of plant tissue in liquid nitrogen and store the samples in a freezer (−80°C). Grind the plant tissue in liquid nitrogen using a mortar and pestle until the tissue becomes a powder. Add 15 ml of grinding buffer (50 mM Tris-HCl, pH 8.0, 0.3 M NaCl, 5 mM EDTA, 2% SDS, 2 mM ATA, 14 mM 2-mercaptoethanol; autoclave before adding ATA and 2-mercaptoethanol) and continue grinding. Pour the slurry into a 50-ml tube containing 2 ml of 3 M KCl. Mix well. Leave the samples on ice for 10 min. Centrifuge the tubes for 5 min at 8,000 rpm at 4°C (Hitachi rotor no. 25 or equivalent). Put the tubes back on ice. Transfer the supernatant through a nylon-mesh filter to a 50-ml tube containing 5 ml of 8 M LiCl. Mix well by inversion, then precipitate overnight at 4°C. Centrifuge for 20 min at 12,000 rpm at 4°C (Hitachi rotor no. 25 or equivalent). Decant the supernatant and put the tubes back on ice. Resuspend the pellet in 2 ml of DEPC-H_2O. Transfer the suspension into a 15-ml tube containing 2 ml of TE-saturated phenol. Vortex the tubes for a few seconds. Centrifuge the tubes for 15 min at 3,500 rpm at 4°C (Hitachi rotor no. 39 or equivalent). Transfer the upper phase into a 15-ml tube containing 200 μl (0.1 vol) of 5 M NaCl and 5 ml (2.5 vol) of cold 99.5% ethanol. Mix well. Store the tubes at −80°C for 1 h. Centrifuge the tubes for 30 min at 3,500 rpm at 4°C (Hitachi rotor no. 39 or equivalent). Decant the supernatant. Rinse the pellets with 2 ml of 70% ethanol. Centrifuge the tubes for 5 min at 3,600 rpm at 4°C (Hitachi rotor no. 39 or equivalent). Decant supernatant. Vacuum-dry the pellets. Suspend the pellet in 50–500 μl of DEPC-H_2O. Measure the RNA concentration using a spectrophotometer. Store the samples in at −80°C.

4.6.3 RNA Blotting

The RNA is electrophoresed through a 1% agarose gel (15 × 20 cm) containing formaldehyde (Caution: Formaldehyde vapors are toxic. Solutions containing formaldehyde should be prepared in a chemical fume hood, and electrophoresis tanks containing formaldehyde solutions should be kept covered whenever possible). Thirty samples (15 lanes × 2 columns) can be examined in a gel.

Place 2 g of agarose (type II, Sigma) and 150 ml of DEPC-H_2O in a 500-ml flask. After dissolving the agarose completely in a microwave oven (~5 min), cool the solution to 60–70°C and add 40 ml of 5 × FGRB (5× formaldehyde gel-running buffer; 0.1 M MOPS, pH 7.0, 40 mM sodium acetate, 5 mM EDTA; sterilize the solution by filtration using a 0.45-μm-pore filter) and 10 ml of formaldehyde in a chemical fume hood. Fill up to 200 ml with DEPC-H_2O and mix well. Ethidium bromide (EtBr) can be added to the gel, but prolonged staining with EtBr is not recommended before RNA is transferred from agarose gels to membrane filters because saturation of the nucleic acid with the dye appears to reduce the efficiency of transfer. Pour the gel in a horizontal gel maker and allow it to set for at least 1 h in the hood. Electrophoresis tanks used for electrophoresis of RNA should be cleaned with detergent solution and rinsed with water. The tanks should then be rinsed thoroughly with DEPC-H_2O.

While incubating the gel in the fume hood, prepare the RNA samples for electrophoresis. In sterile micro-centrifuge tubes, mix 9 μl of RNA sample (10–40 μg of RNA adjusted to 9 μl with DEPC-H_2O), 4 μl of 5× FGRB, 7 μl formaldehyde, 4 μl of BPB solution (50% glycerol, 1 mM EDTA, 0.25% bromophenol blue), and 20 μl deionized formamide. Formamide should be deionized by adding Dowex XG8 mixed-bed resin, stirring on a magnetic stirrer for 1 h, and filtering twice through Whatman no. 1 paper; store in small aliquots at −80°C. Close the tops of the micro-centrifuge tubes and incubate the RNA solutions for 15 min at 65°C. Chill the samples in ice water for 5 min and then centrifuge them for 5 s to deposit all of the fluid in the bottoms of the tubes.

Put the agarose gel in the electrophoresis tank with the running buffer (5 × dilution of 5 × FGRB with DEPC-H_2O). Before loading the samples, pre-run the gel for 5 min at 5 V/cm. Immediately load the RNA samples into the wells of the gel. If possible, leave the first and last lanes of the gel empty. Run the gel submerged in the running buffer for 3.5 h at 80 V constant (3–4 V/cm). After 30 min of electrophoresis, recirculate the running buffer using magnetic stirrers. Constant recirculation of the solution is required to maintain the pH.

At the end of run (when the BPB has migrated approximately 8 cm), stop the electrophoresis, and transfer the RNA from the gel to a nylon membrane by a capillary method. Place a piece of Whatman 3MM paper on a piece of Plexiglas or a stack of glass plates to form a support that is longer and wider than the gel. Place the support inside a large baking dish. Fill the dish with 20× SSC until the level of the liquid reaches almost to the top of the support. When the 3MM paper on the top of the support is thoroughly wet, smooth out all air bubbles with a glass rod. Cut a piece of nylon membrane (Amersham Pharmacia Hybond-XL, or equivalent) about 1 mm larger than the gel in both dimensions. Use gloves and blunt-ended forceps (e.g. Millipore forceps) to handle the filter. Float the nylon membrane filter on the surface of a dish of deionized water until it wets completely from beneath, and then immerse the filter in 20× SSC for 5 min. If the filter is not saturated after floating for several minutes on water, it should be replaced with a new membrane, since the transfer of RNA to an unevenly wetted filter is unreliable. Place the gel on the support in an inverted position so that it is centered on the wet 3MM paper. Make sure that there are no air bubbles between the 3MM paper and the gel. Surround, but do not cover, the gel with Saran Wrap or Parafilm. This serves as a barrier to prevent liquid from flowing directly from the reservoir to paper towels placed on the top of the gel. If these towels are not precisely stacked, they tend to droop over the edge of the gel and may touch the support. This type of short-circuiting is a major reason for insufficient transfer of RNA from gel to the membrane. Place the wet nylon membrane on the top of the gel so that the corners are almost aligned. Do not move the filter once it has been applied to the surface of the gel. Make sure that there are no bubbles between the filter and the gel. Wet two pieces of 3MM paper (cut to exactly the same size as the gel) in 2× SSC and place them on the top of the wet nylon membrane filter. Smooth out any air bubbles with a glass rod. Cut a stack of paper towels (5–8 cm high) just smaller than the sheets of 3MM paper. Place the towels on the sheets of 3MM paper. Put a glass plate on the top of the stack and weigh it down with a 500-g weight. The objective is to set up a flow of liquid from the reservoir through the gel and the nylon membrane filter, so that RNA molecules are eluted from the gel and are deposited on the nylon membrane filter. Allow transfer of RNA to proceed for 6–18 h. As the paper towels become wet, they should be replaced. Remove the paper towels and 3MM papers above the gel. Turn over the gel and the nylon membrane filter and lay them, gel side up, on a dry sheet of 3MM paper. Mark the positions of the gel slots on the filter with a very-soft-lead pencil or a ballpoint pen. Peel the gel from the filter and discard it. Soak the filter in 2× SSC for 1 min at room temperature to remove any agarose sticking to the filter. Remove the filter from the 2× SSC and allow excess fluid to drain away. Place the filter flat on a 3MM paper to dry for 1 h at room temperature. Place the dried filter between two pieces of 3MM paper, and bake the filter for 2 h at 80°C in a vacuum oven. If the membrane filter is not to be used immediately in hybridization experiments, it should be packed in a vinyl hybridization bag and stored at −20°C.

4.6.4 Northern Hybridization

Prehybridize the membrane filter overnight in a hybridization bag or the tube of Hybridiser (HB-1D Hybridiser, Techne, UK, or equivalent) containing a hybridization buffer (Modified Church and Gilbert Buffer: 0.5 M phosphate buffer, pH 7.2, 7% (w/v) SDS, 10 mM EDTA). Radiolabel 25–100 ng of the GUS gene fragment and/or the gene probe of interest with ^{32}P using a random DNA labeling kit (Boehringer Mannheim labeling kit or equivalent). Remove the unlabeled ^{32}P using a spin-column containing Sepharose G-50 with TE (Sigma, Eppendorf). Estimate the radioactivity using a quick counter or a scintillation counter. Denature the probe (heat the DNA at 95°C for 2 min, then chill on ice for 2 min). Add the denatured radiolabeled probe directly to the prehybridization fluid. To detect low-abundance mRNA molecules, use at least 0.1 μg of a probe whose specific activity exceeds 2×10^8 cpm/μg. Continue incubation for 16–24 h at 42°C. Wash the filter for 2×3 min at room temperature in 1× SSC, 0.1% SDS, followed by two washes of 15 min each at 65°C in 0.1× SSC, 0.1% SDS. Place the filter flat on a sheet of 3MM paper to dry for 1 h at room temperature. Keep the blot moist if it is to be reprobed. Cover the filter with Saran Wrap and visualize the hybridized RNA by exposing the filter for 8–24 h to an imaging plate (IP; Fuji BAS2000 system or equivalent) or to X-ray film (Kodak X-ray film or equivalent) at –80°C with an intensifying screen for 24 h to 1 week.

RNA may be stained on the filter after hybridization and exposure to X-ray film. Soak the dried filter in 5% acetic acid for 15 min at room temperature. Transfer the filter to a solution of 0.5 M sodium acetate (pH 5.2) and 0.04% methylene blue for 5–10 min at room temperature. Rinse the filter in water for 10 min to overnight. Using this method, the rRNA bands on the membrane filter can be seen.

4.7 Application of the GUS System

In this chapter, a basic protocol for GUS assay of promoter analysis during dehydration and high-salt responses in plants has been provided. Since *GUS* (as well as *LUC* and *GFP*) is an excellent reporter gene, the GUS system can be applied to several kinds of experiments. Here, we briefly show some examples. For more detailed information about these methods, read the original paper or see other protocols for GUS.

4.7.1 Transient Assay

The promoter-*GUS* reporter system can be used to study expression of the gene in a transient assay. In this case, we use a second reporter gene linked to a constitutively expressed promoter, such as the CaMV35S promoter, as an internal standard. This makes it possible to correct for variations in transformation efficiencies between the samples. For this reason, the promoter-*GUS* and the constitutive promoter-*LUC* constructs are routinely used together in transient assays when the DNA is introduced by the PEG method, electroporation, or microprojective bombardment into leaves or protoplasts (Twell et al. 1989). Both the GUS and the LUC activity are measured, with the ratio of GUS activity to LUC activity reflecting promoter activity. Although the transient assay is not suitable in analyzing dehydration- or salt-induced gene expression, this assay is often used in studying ABA-induced expression (Uno et al. 2000). For the GUS fluorometric assay in the transient assay using protoplasts, following the transformation, thorough washing must be carried out to remove contamination by cell-wall-degrading enzymes, which have GUS-like activity and can interfere with the assay (Topping and Lindsey 1997).

4.7.2 Transactivation Experiment

When isolating *trans*-acting factors, the transactivation activity of the proteins must be assayed. We usually make three kinds of constructs. The first one is the effector construct: constitutive promoter-*TF* (*trans*-acting factor) gene. The second one is the reporter construct: promoter-*GUS* gene. The third one is the control (standard) construct: constitutive promoter-*LUC* gene. All of these constructs are introduced into protoplasts or leaves in a transient assay. In addition, a control effector construct is introduced t instead of the effector construct containing the *TF* gene. The GUS and LUC activities are then measured and the ratio of GUS activity to LUC activity determined. The transactivation activity is revealed by comparing the data from the ratio with the *TF* gene with the data from the ratio without the *TF* gene. For more information on this kind of experiment, please see the original papers (Abe et al. 1997; Liu et al. 1998; Uno et al. 2000).

4.7.3 Promoter Tagging (Enhancer Trap)

The *GUS* reporter gene is used in a program of insertional mutagenesis by transformation with a promoter trap vector to identify genes involved in plant development (see Topping et al. 1994 for more detail). This technique depends on the "positional effect" of the transformation: the insertion of the T-DNA or transposon occurs randomly in the genome. This system can be used to identify stress-inducible promoters in the genome.

4.8 Conclusion

Promoter analysis using the GUS system can open up new areas of study. After identification of the DRE *cis*-element, important in gene expression in dehydration, high salt, and cold stress, we succeeded in isolating the *trans*-acting factors that bind to the DRE element and activate transcription. Two types of DRE-binding proteins (DREB1A and DREB2A) were isolated in *Arabidopsis* using yeast one-hybrid screening with a cDNA library prepared from dehy-

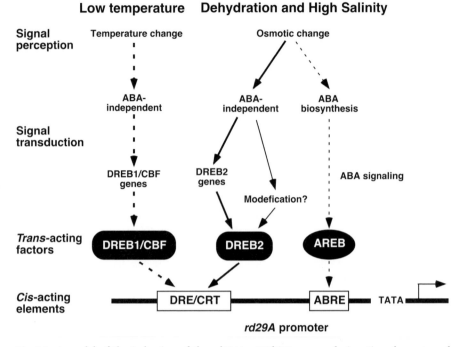

Fig. 4.8. A model of the induction of the *rd29A/cor78/lti78* gene and *cis*-acting elements and *trans*-acting factors involved in stress-responsive gene expression (adapted from Shinozaki and Yamaguchi-Shinozaki 2000). Two *cis*-acting elements, DRE/CRT and ABRE, are involved in the ABA-independent and ABA-responsive induction of *rd29A*, respectively. Two different DRE/CRT-binding proteins, DREB1/CBF1 and DREB2, distinguish two different signal transduction pathways in response to cold and dehydration stresses, respectively. DRE/CRT-binding proteins contain an AP2 (apetala 2) DNA-binding domain, whereas ABRE-binding proteins (*AREB*) encode bZIP transcription factors. *Thick broken arrows* represent a cold signaling pathway. *Solid thick arrows* and *thin broken arrows* represent ABA-independent and an ABA-dependent signaling pathways, respectively, that are involved in the dehydration response

drated or cold-treated *Arabidopsis* plants (Fig. 4.8; Liu et al. 1998). The deduced amino acid sequences of *DREB1A* and *DREB2A* showed no significant sequence similarity except in the conserved DNA-binding domain found in the ERF (ethylene responsive factor) and AP2 (apetala 2) proteins. DREB1A and DREB2A specifically bound to the DRE sequence in vitro and activated transcription of the *GUS* reporter gene driven by the DRE sequence in *Arabidopsis* protoplasts. In these transactivation experiments the *GUS* and *LUC* reporter genes were used (Liu et al. 1998). Expression of the *DREB1A* gene was induced by cold stress, and expression of the *DREB2A* gene was induced by dehydration and high salinity. Two independent DREB proteins, DREB1A and DREB2A, function as transcriptional activators in two separate signal transduction pathways under low temperature and dehydration conditions, respectively (Liu et al. 1998). Subsequently, two homologues of DREB1A (DREB1B and DREB1C) and one homologue of DREB2A (DREB2B) were identified in *Arabidopsis* (Shinwari et al. 1998, Nakashima et al. 2000). Northern analysis and promoter-*GUS* experiments showed that all genes in the DREB1 family are expressed under cold conditions but not under dehydration and high salt conditions (Shinwari et al. 1998). On the other hand, the DREB2B gene is upregulated under dehydration and high salt conditions but not under cold conditions (Nakashima et al. 2000). Overexpression of the DREB1A cDNA in transgenic *Arabidopsis* plants induced strong expression of its target genes under unstressed control conditions. The transgenic plants that overexpressed the DREB1A cDNA revealed tolerance not only to freezing but also to dehydration (Liu et al. 1998; Kasuga et al. 1999). Thomashow's group also conducted similar experiments, almost simultaneously with ours, and reported similar results using CBF1 (DREB1B) and CBF3 (DREB1A) in low-temperature stress (Stockinger et al. 1997; Jaglo-Ottosen et al. 1998; Gilmour et al. 1998, 2000). Thus promoter analysis using the promoter-*GUS* reporter system can further research not only in a basic biological direction but also in an applied biological direction.

Sequencing of the *Arabidopsis* genome was completed at the end of the year 2000, and the structure of all 25,000 *Arabidopsis* genes was determined (Nature, 14 December 2000). Recently, sequencing of the rice genome has also been completed by two companies, Syngenta (Basel, Switzerland) and Myriad Genetics (Salt Lake City, UT). All stress-inducible genes can then be identified by the systemic analysis of gene expression in microarrays. Using microarray experiments, we showed that many unidentified genes are upregulated under dehydration, high salt and cold conditions in *Arabidopsis* (Seki et al. 2001). For the first time, genomic studies and microarray analysis will facilitate an overall understanding of the molecular networks responsive for stress responses in plants. In parallel with this, detailed expression analysis of individual genes using the GUS system will be retained as an important tool in investigations of plant stress responses at the molecular level.

Acknowledgements. We thank Dr. Sean D. Simpson, JIRCAS, for a critical reading of the manuscript.

References

Abe H, Yanaguchi-Shinozaki K, Urao T, Iwasaki T, Hosokawa D, Shinozaki K (1997) Role of *Arabidopsis* MYC and MYC homologs in drought- and abscisic acid-regulated gene expression. Plant Cell 9:1859–1868

Bechtold N, Pelletie G (1998) In planta *Agrobacterium*-mediated transformation of adult *Arabidopsis thaliana* plants by vacuum infiltration. In: Martines-Zapater J, Salinas J (eds) Methods in molecular biology 82, *Arabidopsis* protocols. Humana Press, Totowa, NJ, pp 259–266

Bray EA (1997) Plant responses to water deficit. Trends Plant Sci 2:48–54

Clough SJ, Bent AF (1998) Floral dip: a simplified method for Agrobacterium-mediated transformation of *Arabidopsis thaliana*. Plant J 16:735–743

De Block M, Van Lijsebettens M (1998) β-Glucuronidase enzyme histochemistry on semithin sections of plastic-embedded *Arabidopsis* explants. In: Martines-Zapater J, Salinas J (eds) Methods in molecular biology 82, *Arabidopsis* protocols. Humana Press, Totowa, NJ, pp 397–407

Gallagher SR (ed) (1992) GUS protocols: using the GUS gene as a reporter of gene expression. Academic, San Diego, pp 1–221

Gilmour SJ, Zarka DG, Stockinger EJ, Salazar MP, Houghton JM, Thomashow MF (1998) Low temperature regulation of the *Arabidopsis* CBF family of AP2 transcriptional activators as an early step in cold-induced COR gene expression. Plant J 16:433–443

Gilmour SJ, Sebolt AM, Salazar MP, Everard JD, Thomashow MF (2000) Overexpression of the *Arabidopsis* CBF3 transcriptional activator mimics multiple biochemical changes associated with cold acclimation. Plant Physiol 124:1854–1865

Hasegawa PM, Bressan RA, Zhu J-K, Bohnert HJ (2000) Plant cellular and molecular responses to high salinity. Annu Rev Plant Physiol Plant Mol Biol 51:463–499

Hull GA, Devic M (1995) The β-glucuronidase (gus) reporter gene system. In: Jones H (ed) Methods in molecular biology 47. Plant gene transfer and expression protocols. Humana, Totowa, NJ, pp 125–141

Ingram J, Bartels D (1996) The molecular basis of dehydration tolerance in plants. Annu Rev Plant Physiol Plant Mol Biol 47:377–403

Ishitani M, Xiong L, Stevenson B, Zhu J-K (1997) Genetic analysis of osmotic and cold stress signal transduction in *Arabidopsis*: interactions and convergence of abscisic acid-dependent and abscisic acid-independent pathways. Plant Cell 9:1935–1949

Iturriaga G, Jefferson RA, Bevan MW (1989) Endoplasmic reticulum targeting and glycosylation of hybrid proteins in transgenic tobacco. Plant Cell 1:381–390

Jaglo-Ottosen KR, Gilmour SJ, Zarka DG, Schabenberger O, Thomashow MF (1998) *Arabidopsis* CBF1 overexpression induces cor genes and enhances freezing tolerance. Science 280:104–106

Jefferson RA (1988) Plant reporter genes: the *gus* gene fusion system. Genet Eng 10:247–263

Jefferson RA, Wilson KJ (1991) The GUS gene fusion system. Plant Mol Biol Man B14:1–33

Jefferson RA, Burgess SM, Hirsh D (1986) β-Glucuronidase from *Escherichia coli* as a gene-fusion marker. Proc Natl Acad Sci USA 83:8447–8451

Jefferson RA, Kavanagh TA, Bevan MW (1987) GUS fusions: β-glucuronidase as a sensitive and versatile gene fusion marker in higher plants. EMBO J 6:3901–3907

Kasuga M, Liu Q, Miura S, Yamaguchi-Shinozaki K, Shinozaki K (1999) Improving plant drought, salt, and freezing tolerance by gene transfer of a single stress-inducible transcription factor. Nat Biotechnol 17:287–291

Liu Q, Sakuma Y, Abe H, Kasuga M, Miura S, Yamaguchi-Shinozaki K, Shinozaki K (1998) Two transcription factors, DREB1 and DREB2, with an EREBP/AP2 DNA binding domain, separate two cellular signal transduction pathways in drought- and low temperature-responsive gene expression, respectively, in *Arabidopsis*. Plant Cell 10:1391–1406

Millar AJ, Short SR, Hiratsuka K, Chua N-H, Kay SA (1992) Firefly luciferase as a reporter of regulated gene expression in higher plants. Plant Mol Biol Reporter 10:324–337

Nakashima K, Kiyosue T, Yamaguchi-Shinozaki K, Shinozaki K (1997) A nuclear gene, erd1, encoding a chloroplast-targeted Clp protease regulatory subunit homolog is not only induced by water stress but also developmentally upregulated during senescence in *Arabidopsis thaliana*. Plant J 12:851–862

Nakashima K, Satoh R, Kiyosue T, Yamaguchi-Shinozaki K, Shinozaki K (1998) A gene encoding proline dehydrogenase is not only induced by proline and hypo-osmolarity but is also developmentally regulated in reproductive organs in *Arabidopsis thaliana*. Plant Physiol 118:1233–1241

Nakashima K, Shinwari ZK, Sakuma Y, Seki M, Miura S, Shinozaki K, Yamaguchi-Shinozaki K (2000) Organization and expression of two *Arabidopsis* DREB2 genes encoding DRE-binding proteins involved in dehydration- and high-salinity-responsive gene expression. Plant Mol Biol 42:657–665

Sambrook J, Russell DW (2001) Molecular cloning, a laboratory manual, 3rd edn. Cold Spring Harbor Laboratory, Cold Spring Harbor, New York

Seki M, Narusaka M, Abe H, Kasuga M, Yamaguchi-Shinozaki K, Carninci P, Hayashizaki Y, Shinozaki K (2001) Monitoring expression pattern of 1300 *Arabidopsis* gene under dehydration and cold stress using full-length cDNA microarray. Plant Cell 13:61–72

Sheen J, Hwang S, Niwa Y, Kobayashi H, Galbraith DW (1995) Green-fluorescent protein as a new vital marker in plant cells. Plant J 8:777–784

Shinozaki K, Yanaguchi-Shinozaki K (1997) Gene expression and signal transduction in water-stress response. Plant Physiol 115:327–334

Shinozaki K, Yanaguchi-Shinozaki K (1999) Molecular responses to drought stress. In: Shinozaki K, Yanaguchi-Shinozaki K (eds) Molecular responses to cold, drought, heat and salt stress in higher plants. RG Landes, Austin, TX, pp 11–28

Shinozaki K, Yanaguchi-Shinozaki K (2000) Molecular responses to dehydration and low temperature: differences and cross-talk between two stress signaling pathways. Curr Opin Plant Biol 3:217–223

Shinwari ZK, Nakashima K, Miura S, Seki M, Yamaguchi-Shinozaki K, Shinozaki K (1998) An *Arabidopsis* gene family encoding DRE/CRT binding proteins involved in low-temperature-responsive gene expression. Biochem Biophys Res Comm 250:161–170

Stockinger EJ, Gilmour SJ, Thomashow MF (1997) *Arabidopsis* thaliana CBF1 encodes an AP2 domain-containing transcription activator that binds to the C-repeat/DRE, a *cis*-acting DNA regulatory element that stimulates transcription in response to low temperature and water deficit. Proc Natl Acad Sci USA 94:1035–1040

Topping JF, Lindsey K (1997) Molecular characterization of transformed plants. In: Clark MS (ed) Plant molecular biology-a laboratory manual. Springer, Berlin Heidelberg New York, pp 427–442

Topping JF, Agyeman F, Henricot B, Lindsey K (1994) Identification of molecular markers of embryogenesis in *Arabidopsis thaliana* by promoter trapping. Plant J 5:895–903

Twell D, Klein TM, Fromm ME, Mccormick S (1989) Transient expression of chimeric genes delivered into pollen by microinjectile bombardment. Plant Physiol 91:1270–1274

Uno Y, Furihata T, Abe H, Yoshida R, Shinozaki K, Yamaguchi-Shinozaki K (2000) *Arabidopsis* basic leucine zipper transcription factors involved in a abscisic acid-dependent signal transduction pathway under drought and high-salinity conditions. Proc Natl Acad Sci USA 97: 11632–11637

Valvekens D, VanMontagu M, Van Lijsebettens M (1988) *Agrobacterium tumefaciens*-mediated transformation of *Arabidopsis thaliana* root explants by using kanamycin selection. Proc Natl Acad Sci USA 85:5536–5540

Weigel D, Ahn JH, Blazquez MA, Borevitz JO, Christensen SK, Fankhauser C, Ferrandiz C, Kardailsky I, Malancharuvil EJ, Neff MM, Nguyen JT, Sato S, Wang Z-Y, Xia Y, Dixon RA, Harrison MJ, Lamb CJ, Yanofsky M, Chory J (2000) Activation tagging in *Arabidopsis*. Plant Physiol 122:1003–1013

Wilkinson JE, Twell D, Lindsey K (1994) Methanol does not specially inhibit endogenous β-glucuronidase (GUS) activity. Plant Sci 97:61–67

Yamaguchi-Shinozaki K, Shinozaki K (1994) A novel cis-acting element in an *Arabidopsis* gene is involved in responsiveness to drought, low-temperature, or high-salt stress. Plant Cell 6:251–264

Yoshiba Y, Nanjo T, Miura S, Yamaguchi-Shinozaki K, Shinozaki K (1999) Biochem Biophys Res Comm 261:766–772

5 Methods for Detecting Genetic Manipulation in Grain Legumes

H.-J. JACOBSEN and R. GREINER

5.1 Introduction

Novel foods have been introduced throughout mankind's history, and they always have enriched, sometimes boring, diets. Nobody in Europe could imagine times before plants like potatoes, tomatoes, corn, squash and melons, beans and many others species that were introduced from foreign countries and continents. Nowadays, especially in highly developed countries, where hunger is seldom and surpluses of food are frequent, novel foods are seen rather critically, and related reports in the news often create doubts and fears, particularly with food derived from genetically engineered crop plants. In the opinion of many consumers consumption of genetically manipulated (GM) foods implies unintended risks for human health. Furthermore, environmental and economic concerns as well as religious and ethical considerations have been raised. However, while the risks are often stressed, it has not been mentioned how unlikely it is that such scenarios will come true and that some of the scenarios are only hypothetical. While some safety concerns have been raised regarding the introduction of GM foods to the market, these concerns should be viewed as a normal consequence of the introduction of a dramatically new technology. The food industry has a life-long tradition regarding the introduction of novel foods. Therefore, it is clear that it is anything but novel for the food industry to responsibly introduce these foods. Processing and production of foods are constantly changing, and today not only wild-type species, but also organisms (microorganisms, plants, animals) improved by traditional breeding are used. To give consumers a choice whether or not to buy GM foods, their identification and labeling are required. These detection methods are also important, e.g., for trade with grains with respect to guaranteeing their authenticity. Regulations such as the Novel Food Directive (Regulation 1997) and follow-up directives (Regulation 1998, 2000a, b) within the European Union are likely to come into force in Europe, requiring the identification of crop plants, processed foods and animal feed derived from genetic engineering with unforeseen consequences for the rest of the world. The German working group "Development of methods to identify foods produced by means of genetic engineering" has already carried out interlaboratory studies on several products derived from genetic engineering including soybeans (Jankiewicz et al. 1999), which today are of particular importance in the grain legume area, but other grain legumes will follow once GM varieties reach the

Molecular Methods of Plant Analysis, Vol. 22
Testing for Genetic Manipulation in Plants
Edited by J.F. Jackson, H.F. Linskens, and R.B. Inman
© Springer-Verlag Berlin Heidelberg 2002

market. The group successfully applied for the inclusion of those methods into the Official Collection of Methods according to LMBG §35 of the German Food Act (Jankiewicz et al. 1999). Detection methods for products derived from genetic engineering have mainly focused on the newly introduced DNA (Gachet et al. 1999; Lüthy 1999; Meyer 1999), but the newly introduced protein (Brett et al. 1999; Rogan et al. 1999; Lipton et al. 2000) or any other difference in composition such as fatty acid composition of a vegetable oil can also be used for detection of GM food.

5.2 Detection at the DNA Level

Because of its high sensitivity, specificity, and rapidity, the polymerase chain reaction (PCR) is the method of choice for detecting transgenic sequences in crop plants, processed foods, and animal feed (Meyer 1995, 1999; Koeppel et al. 1997; Konietzny and Greiner 1997; Hoef et al. 1998; Gachet et al. 1999; Jankiewicz et al. 1999; Lipp et al. 1999; Lüthy 1999; Van Duijn et al. 1999; Greiner and Jany 2001). Prerequisites for the use of PCR to detect transgenic sequences are:

- Knowledge of the genetic modification
- Presence of DNA in the product to be analyzed
- Extractability of DNA in sufficient quantity, purity, and fragment length for PCR analysis from the product to be analyzed

Knowledge of the genetic modification is required to design specific PCR primers. This is a critical step in PCR analysis, since the PCR primers need to have the required sensitivity and specificity. It was shown that the sensitivity of a certain PCR may significantly differ from laboratory to laboratory and from matrix to matrix (Lipp et al. 1999). This may result in different conclusions regarding the sample material. Therefore, a uniform reference material has to be made available in order to adjust the parameters of the methodology in a way that allows the detection of a certain relative percentage of GM material (1% within the European Union) in a product.

The presence of DNA in the sample to be analyzed depends on the history of the product. DNA is always available if the product is the genetically engineered organism itself or if the product contains genetically engineered organisms. However, processing is one of the main factors negatively influencing the accessibility of appropriate nucleic acid substrate for PCR. During processing, fragmentation of DNA could occur, for example, by mechanical treatment (shear-forces), enzymatic reaction (nucleases), or chemical hydrolysis (acidic pH). Moreover, processing may lead to complete degradation or removal of the DNA. Thus, the presence of suitable DNA for PCR depends on the product itself, the production process, and the storage parameters. The possibility of identifying a processed product as produced through genetic engineering has therefore to be elucidated on a case-by-case basis (Greiner et al. 1997; Pauli et al. 1998). However, even products that do not contain DNA could be identified as derived from genetic engineering by using a certification program that employs identity preservation and traceability systems, along with testing at earlier stages of processing, when extractable DNA still is present.

5.3 PCR Analysis

PCR analysis can be divided into the following steps:

1. DNA extraction
2. Amplification of target sequences by PCR
3. Detection of amplification products by agarose gel electrophoresis
4. Verification of the amplification product

The amount and quality of the extracted DNA depend largely on the combination of sample matrix and extraction method used (Lick et al. 1996; Zimmermann et al. 1998; Tengel et al. 2000). The classic *N*-cetyl-*N,N,N*-trimethylammoniumbromide (CTAB)-based methods are widely used to extract DNA from plant-based material (Jankiewicz et al. 1999):

1. Add 1 ml of 100 mM Tris, 20 mM Na_2-EDTA, 1.4 M NaCl, 20 g/l CTAB, pH 8.0, to 100 mg lyophilized sample material in a 2-ml reaction tube and mix thoroughly.
2. Incubate for 30 min at 65°C.
3. Centrifuge for 10 min at 10,000 g.
4. Transfer 500 µl of the supernatant to a new reaction tube, add 200 µl chloroform and mix gently.
5. Centrifuge for 10 min at 10,000 g.
6. Transfer the upper phase into a new reaction tube, add 2 vol of 40 mM NaCl, 5 g/l CTAB, and mix gently.
7. Incubate for 60 min at room temperature.
8. Centrifuge for 5 min at 10,000 g
9. Discard the supernatant and dissolve the precipitate in 350 µl 1.2 M NaCl.
10. Add 350 µl chloroform and mix gently.
11. Centrifuge for 10 min at 10,000 g.
12. Transfer the upper phase into a new reaction tube, add 0.6 volumes of isopropanol and mix gently.
13. Centrifuge for 10 min at 10,000 g.
14. Discard the supernatant, add 500 ml ethanol (70%) to the pellet and mix gently.
15. Centrifuge for 10 min at 10,000 g.
16. Discard the supernatant, air-dry the precipitate and dissolve the precipitate in 10 mM Tris, 1 mM Na_2-EDTA, pH 8.0.

The above described CTAB-based method results for a multitude of plant-based material in the extraction of sufficient DNA suitable for PCR. For several matrices such as flour, protein isolates or non-processed fruits, steps 5–10 could be omitted without significantly effecting PCR. If the samples contain high amounts of oil or fat, such as chocolate, extracting the samples with hexane prior to the above described DNA extraction results, in general, in a DNA preparation suitable for PCR. Currently, several DNA isolation kits developed for DNA isolation from plant-based material such as food or feed are commercially available. In general, the CTAB-based methods result in higher DNA yields but poorer DNA quality, and the extraction is more time-consuming compared to the use of kits (Greiner and Jany 2001).

Because of the existence of a great number of very different products which have to be analyzed, the yield and purity of the extracted DNA could be improved by adapting the extraction procedure to the matrix of the DNA source. This is especially important since a number of accompanying compounds such as fat or protein are potent PCR inhibitors (Rossen et al. 1992; Dickinson et al. 1995). The exclusion of PCR inhibitors is a very crucial point in PCR analysis. Unfortunately, there are only a very limited number of investigations on plant and food components inhibiting PCR. To exclude false-negative results, the absence of inhibitors has to be shown by a control PCR (see below) using an internal standard or by spiking the DNA preparation with the target sequence.

5.4 Control PCR and Specific PCR Systems

To identify a product as being derived from genetic engineering by PCR, it has to be shown first that purity and yield of the extracted DNA are sufficient for PCR. The presence of amplifiable DNA can be determined by PCR using a target sequence always present in the product to be analyzed. For that purpose 18S RNA (eukaryote-specific) or the soybean lectin gene (soybean-specific) could be used as target sequences (Table 5.1).

In addition to the control PCR, a specific detection system is needed to identify a product as derived from genetic engineering. Since most of the transgenic crops approved contain the ^{35}S promoter of cauliflower mosaic virus and/or the NOS terminator of *Agrobacterium*, these genetic elements have been used as target sequences for general screening (Pietsch et al. 1997; Wolf et al. 2000; Table 5.2). This approach has the advantage of being able to confirm the presence of transgenic sequences from a wide number of different GM organisms (GMO). The absence of an amplification product points to the absence of transgenic sequences, whereas the presence of an amplification product points to the presence of a GMO or contamination by cauliflower mosaic virus or *Agrobacterium*. Since it is not possible to identify the GMO itself by using the screening approach, a unique DNA sequence for the GMO to be identified has

Table 5.1. Control PCR for the RoundupReady soybean. The PCR mixture consisted of: 17.4 µl H_2O, 2.5 µl PCR buffer (10×), 2 µl dNTP solution (2.5 mM each), 1 µl of each primer (5 µM), 1.0 µl DNA solution, 0.1 µl AmpliTaqGold (5 U/µl)

Target sequence	Forward primer/ reverse primer	Fragment length of the amplification product (bp)	PCR program	Reference
18S RNA	TR03: 5'-TCT gCC CTA TCA ACT TTC gAT ggT A-3'	137	10 min at 95 °C	Allmann et al. (1993)
	TR04: 5'-AAT TTg CgC gCC TgC TgC Ctt CCT T-3'		35 cycles: 30 s at 95 °C, 60 s at 68 °C, 60 s at 72 °C 3 min at 72 °C	
Soybean lectin	GM03: 5'-gCC CTC TAC TCC ACC CCC ATC C-3'	118	10 min at 95 °C	Jankiewicz et al. (1999)
	GM04: 5'-gCC CAT CTg CAA gCC TTT TTg Tg-3'		35 cycles: 30 s at 95 °C, 30 s at 60 °C, 60 s at 72 °C 3 min at 72 °C	

Table 5.2. Screening method. The PCR mixture consisted of: 17.4 µl H_2O, 2.5 µl PCR buffer (10×), 2 µl dNTP solution (2.5 mM each), 1 µl of each primer (5 µM), 1.0 µl DNA solution, 0.1 µl Ampli-TaqGold (5 U/µl). The amplification products could be verified by restriction analysis: digestion of the 35S-1/35S-2 amplification product with *Xmn*I results in two fragments, 115 and 80 bp, and digestion of the NOS-1/NOS-2 amplification product with *Nsi*I result in two fragments, 96 and 84 bp. *CaMV* Cauliflower mosaic virus

Target sequence	Forward primer/ reverse primer	Fragment length of the amplification product (bp)	PCR program	Reference (19976)
CaMV ^{35}S promotor	35S-1: 5'-gCT CCT ACA AAT gCC ATC A-3'	195	10 min at 95 °C	Pietsch et al.
	35S-2: 5'-gCC CAT CTg CAA gCC TTT TTg Tg-3'		35 cycles: 20 s at 95 °C, 40 s at 54 °C, 40 s at 72 °C 3 min at 72 °C	
Agrobacterium NOS terminator	NOS-1: 5'-gAA TCC TgT TgC Cgg TCT Tg-3'	180	10 min at 95 °C	Pietsch et al. (1997)
	NOS-3: 5'-TTA TCC Tag TTT gCg CgC TA-3'		35 cycles: 20 s at 95 °C, 40 s at 54 °C, 40 s at 72 °C 3 min at 72 °C	

Table 5.3. Specific detection system for the RoundupReady soybean. The PCR mixture consisted of: 17.4 μl H$_2$O, 2.5 μl PCR buffer (10×), 2 μl dNTP solution (2.5 mM each), 1 μl of each primer (5 μM), 1.0 μl DNA solution, 0.1 μl AmpliTaqGold (5 U/μl). The amplification product could be verified by Southern blotting and hybridization; hybridization probe: 5'-ggg TCT TgC gAA ggA Tag Tg-3'. *CaMV* Cauliflower mosaic virus

Target sequence	Forward primer/ reverse primer	Fragment length of the amplification product (bp)	PCR program	Reference
CaMV ^{35}S promoter/ chloroplast transit signal sequence	p35S-f2: 5'-TgA TgT gAT ATC TCC ACT gAC g-3'	172	10 min at 95 °C	Jankiewicz et al. (1999)
	petu-r1: 5'-TgT ATC CCT TgA gCC Atg TTg T-3'		35 cycles: 30 s at 95 °C, 30 s at 62 °C, 25 s at 72 °C 3 min at 72 °C	

to be used as a target sequence in the PCR (Table 5.3). Unique DNA sequences are part of the newly introduced DNA and span the boundary of at least two adjacent genetic elements. To increase the probability of obtaining amplification products even when DNA has to be isolated from highly processed foods, PCR primers should be chosen with the objective to amplify rather small products (100–200 bp; Tables 5.1–5.3).

To exclude false-positive and/or false-negative results in the PCR analysis of crop plant, food or feed, several controls have to be included into the methodology:

– Control of DNA yield and DNA quality (see above)
– Positive control (to check the PCR itself)
– Negative control (to exclude contamination of the PCR reagents)
– Extraction control (extraction procedure without a DNA source to exclude contamination of the laboratory and the extraction reagents)

The amplification products could be analyzed by agarose gel electrophoresis. The transgenic sequence should be detected if the fragment length of the amplification product is of the same size as the target sequence. A verification of the PCR result is highly recommended, because sometimes false-positive results have been observed, i.e., an amplification product not derived from the target sequence but, because of a very similar fragment length, not distinguishable from the amplification product of the target sequence. To verify the PCR results (Tables 5.2, 5.3), restriction analysis (specific cleavage of the amplification product by restriction endonuclease), Southern blotting followed by hybridization (transfer of the amplification product onto a membrane followed by hybridization with a DNA probe specific for the target sequence) or sequencing of the amplification product may be used. Furthermore, it has to be kept in mind that the absence of a specific amplification product does not prove the absence of transgenic sequences in the product analyzed, but only that the transgenic sequence which has been looked for is not present.

5.5 Quantitative Approach

So far, PCR methods have been strictly qualitative. It is hence not possible by qualitative PCR to distinguish a product derived from genetic engineering from any contamination non-intentionally brought in by carry-over effects during breeding, storage, processing and transport. Therefore, quantitative PCR systems are needed. For this purpose, competitive PCR (Studer et al. 1998; Huebner et al. 1999a,b; Pietsch et al. 1999) and real-time PCR systems (Wurz et al. 1999) have been developed. The aim of these quantitative approaches is not an absolute but a relative quantification. This means that not the absolute amount of material derived from genetic engineering will be determined in a product but the percentage of GM material with respect to a certain food component, e.g., the percentage of transgenic soybeans in the whole soybean fraction of the product.

5.6 Competitive PCR

The principle of competitive PCR is a co-amplification of a defined amount of target DNA with an increasing amount of an internal DNA standard (competitor). The competitor is a synthetic DNA fragment very similar to the target sequence but different in fragment length and acts also as an internal amplification control. Both amplification products are therefore distinguishable on an agarose gel. For quantification, the intensities of the amplification products are compared after densitometric gel evaluation. Identical amounts of target DNA and competitor at the beginning of the PCR result in identical intensities of the amplification products on the agarose gel. To determine the relative percentage of GM material in a product, a double-competitive PCR system is recommended. Thereby, the relative percentage of GM material could even be determined in mixed and processed products. Double-competitive PCR means that two independent PCR systems are used, one system specific for the GMO, e.g., for the RoundupReady soybean, and one system specific for the species, e.g., for soybeans in general. The relative percentage of GM material can be calculated by the ratio of these two PCRs, i.e., the ratio of the RoundupReady-specific PCR and the soybean-specific PCR. Competitive PCR is a sensitive method for the determination of GM material in plant crops, food and feed, as shown in several interlaboratory studies. However, since competitive PCR is laborious, real-time PCR-systems are preferred in the quantification of GM material.

5.7 Real-Time PCR Systems

For real-time PCR, the LighCycler- and TaqMan techniques are mainly used. Both systems are suitable to determine the relative percentage of GM material not only in raw, but also in mixed and processed products. The TaqMan technique makes use of the 5'-3' exonuclease activity of the polymerase to generate a template-specific fluorescent signal after hydrolyzing an internal probe during each step of the PCR. The parameter measured is the threshold cycle (CT), in which each reaction overcomes a certain fluorescence level. As with competitive PCR, two independent PCR systems are used, one system specific for the GMO, e.g., for the RoundupReady soybean, and one system specific for the species, e.g., for soybeans in general. The relative percentage of GM material can be calculated by the ratio of these two PCR, i.e.,. the ratio of the RoundupReady-specific PCR and the soybean-specific PCR. For both PCR systems calibration curves are needed to make quantification possible. False-negative results due to PCR inhibitors are excluded by using an internal amplification control in each PCR. The TaqMan technique is a robust and sensitive method for the determination of GM material in plant crops, food and feed, as shown in several interlaboratory studies.

5.8 Concluding Remarks

Today, PCR analysis is the only method capable of discriminating between a genetically modified organism and the wild-type. Processing may reduce the amount of DNA in the raw material and/or result in fragmentation of the DNA. But, different from what has been commonly assumed, this prevents detection in exceptional cases only. The crucial point of PCR analysis of crop plants, food and feed is the exclusion of PCR inhibitors.

Quantitative PCR is sufficiently robust to be effective for virtually all DNA containing matrices and has sufficient sensitivity to distinguish a portion of a product derived from genetic engineering below and above the threshold level of 1% legal in the European Union (Regulation 2000b). The method is well-suited for automation and high throughput of samples, and can be used for raw, processed, and even mixed products.

It has to be kept in mind that the described detection methods are only capable of identifying the product under investigation as being derived from genetic engineering. The safety of the product or the risk related to the product has to be addressed separately.

References

Allmann A, Candrian U, Höfelein C, Lüthy J (1993) Polymerase chain reaction (PCR): a possible alternative to immunochemical methods assuring safety and quality of food. Z Lebensm Unters Forsch 196:248–251

Brett GM, Chambers SJ, Huang L, Morgan MRA (1999) Design and development of immunoassays for detection of proteins. Food Control 10:401–406

Dickinson JH, Kroll RG, Grant KA (1995) The direct application of the polymerase chain reaction to DNA extracted from foods. Lett Appl Microbiol 20:212–216

Gachet E, Martin GG, Vingeneau F, Meyer G (1999) Detection of genetically modified organisms by PCR: a brief review of methodologies available. Trends Food Sci Technol 9:380–388

Greiner R, Jany KD (2001) A sensitive tool to evaluate genotypic contaminations from breeding up to processing foods: applicability to legumes. In: Durqanti M et al. (eds) Proceedings 4th European Conference on Grain Legumes, Cracow, Poland, 8–12 July 2001, AEP, Paris, pp 18-19

Greiner R, Konietzny U, Jany KD (1997) Is there any possibility of detecting the use of genetic engineering in processed foods? Z Ernährungswiss 36:155–160

Hoef AM, Kok EJ, Bouw E. Kuiper HA, Keijer J (1998) Development and application of a selective detection method for genetically modified soy and soy-derived products. Food Addit Contam 15:767–774

Huebner P, Studer E, Lüthy J (1999a) Quantitation of genetically modified organisms in food. Nat Biotechnol 17:1137–1138

Huebner P, Studer E, Lüthy J (1999b) Quantitative competitive PCR for the detection of genetically modified organisms in food. Food Control 10:353–358

Jankiewicz A, Broll H, Zagon J (1999) The official method for the detection of genetically modified soybeans (German Food Act LMBG §35: a semi-quantitative study of sensitivity limits with glyphosate-tolerant soybeans (Roundup Ready) and insect-resistant Bt maize (Maximizer). Eur Food Res Technol 209:77–82

Koeppel E, Stadler M, Lüthy J, Hübner P (1997) Sensitive Nachweismethode für die gentechnisch veränderte Sojabohne "Roundup Ready". Mitt Gebiete Lebensm Hyg 88:164–175

Konietzny U, Greiner R (1997) Model systems to develop detection methods for food derived from genetic engineering. J Food Comp Anal 10:28–35

Lick S, Keller M, Bockelmann W, Heller KJ (1996) Optimized DNA extraction method for starter cultures from yoghurt. Milchwissenschaften 51:183–186

Lipp M, Brodmann P, Pietsch K, Pauwels J, Anklam E (1999) IUPAC collaborative trial study of a method to detect genetically modified soy beans and maize in dried powder. J AOAC Int 82:923–928

Lipton CR, Dautlick JX, Grothaus GD (2000) Guidelines for the validation and use of immuno-assays for determination of introduced proteins in biotechnology enhanced crops and derived food ingredients. Food Agric Immunol 12:153–164

Lüthy J (1999) Detection strategies for food authenticity and genetically modified foods. Food Control 10:359–361

Meyer R (1995) Nachweis gentechnologisch veränderter Pflanzen mittels der Polymerase Kettenreaktion (PCR) am Beispiel der FLAVR SAVR™-Tomate. Z Lebensm Unters Forsch 201:583–586

Meyer R (1999) Development and application of DNA analytical methods for the detection of GMO in food. Food Control 10:391–399

Pauli U, Liniger M, Zimmermann A (1998) Detection of DNA in soybean oil. Z Lebensm Unters Forsch A 207:264–267

Pietsch K, Waiblinger HU, Brodmann P, Wurz A (1997) Screeningverfahren zur Identifizierung "gentechnisch veränderter" pflanzlicher Lebensmittel. Dtsch Lebensm Rundsch 93:35–38

Pietsch K, Bluth A, Wurz A, Waiblinger U (1999) Kompetetive PCR zur Quantifizierung konventioneller und transgener Lebensmittelbestandteile. Dtsch Lebensm Rundsch 95:57

Regulation (1997) Regulation (EC) No. 258/97 of the European Parliament and of the Council of 27 January 1997 concerning novel foods and novel food ingredients. Official J Eur Communities L 43:1–7

Regulation (1998) Regulation (EC) No. 1139/98 of the European Parliament and of the Council of 26 May 1998 concerning the compulsory indication of the labelling of certain foodstuffs produced from genetically modified organisms of particulars other than those provided for the Directive 79/112/EEC. Official J Eur Communities L 159/4:4–7

Regulation (2000a) Regulation (EC) No. 50/2000 of the European Parliament and of the Council of 10 January 2000 concerning the labelling of foodstuffs and food ingredients containing additives and flavourings that have been genetically modified or have been produced from genetically modified organisms. Official J Eur Communities L 6:15–17

Regulation (2000b) Regulation (EC) No 49/2000 of the European Parliament and of the Council of 10 January 2000 amending Council Regulation (EC) No 1139/98 concerning the compulsory indication on the labelling of certain foodstuffs produced from genetically modified organisms of particulars other than those provided for in Directive 79/112/EEC. Official J Eur Communities L 6:13–14

Rogan GJ, Dudin YA, Lee TC, Magin KM, Astwood JD, Bhakta NS, Leach JN, Sanders PR, Fuchs RL (1999) Immunodiagnostic methods for detection of 5-enolpyruvylshikimate-3-phosphate synthase in RoundupReady soybeans. Food Control 10:407–414

Rossen L, Nørskov P, Holmstrøm K, Rasmussen OF (1992) Inhibition of PCR by components of food samples, microbial diagnostic assays and DNA extraction solutions. Int J Food Microbiol 17:37–45

Studer E, Rhyner C, Lüthy J, Hübner P (1998) Quantitative competitive PCR for the detection of genetically modified soybean and maize. Z Lebensm Unters Forsch 207:207–213

Tengel C, Schüßler P, Sprenger-Haußels M, Setzke E, Balles J (2000) Verbesserter Nachweis gentechnisch veränderter Soja- und Maisbestandteile in Kakao-haltigen Lebensmitteln sowie in Lebensmittelzusatzstoffen. Dtsch Lebensm Rundsch 96:129–135

Van Duijn G, van Biert R, Bleeker-Marcelis H, Peppelman H, Hessing M (1999) Detection methods for genetically modified crops. Food Control 10:375–378

Wolf C, Scherzinger M, Wurz A, Pauli U, Huebner P (2000) Detection of cauliflower mosaic virus by the PCR; testing of food components for false-positive 35S-promoter screening results. Eur Food Res Technol 210:367–370

Wurz A, Bluth A, Zeltz P, Pfeiffer C, Willmund R(1999) Quantitative analysis of genetically modified organisms (GMO) in processed food by PCR-based methods. Food Control 10:385–389

Zimmermann A, Lüthy J, Pauli U (1998) Quantitative and qualitative evaluation of nine different extraction methods for nucleic acids on soybean food samples. Z Lebensm Unters Forsch 207:81–90

6 Elimination of Selectable Marker Genes from Transgenic Crops

A.P. Gleave

6.1 Introduction

The first reports of successful plant transformation appeared in the early 1980s. Since that time there has been a steady increase in the number of plant species that can be transformed and there have been improvements to the transformation efficiencies of many plant species. However, as DNA uptake and its integration into the genome remain at a low frequency, an essential step in the transformation process is the ability to select transformed cells from the majority of non-transformed cells. This is usually achieved by the expression of a selectable marker gene, linked to the transgene, and selection of transformed cells for their ability to proliferate in the presence of the selective agent. Under appropriate conditions transgenic plants can be regenerated from these selected cells and thereafter the selectable marker gene is generally superfluous. Selectable marker genes have traditionally been antibiotic or herbicide resistance genes, the most commonly used being the neomycin phosphotransferase (*nptII*), hygromycin phosphotransferase (*hyg*) and phosphinotricin acetyl transferase (*bar*) genes which confer resistance to the antibiotics kanamycin, hygromycin and the herbicide glufosinate, respectively.

In recent years, increases in the field release of transgenic plants and the commercialization of their products have led to concerns regarding the effects of the presence of selectable marker genes on ecosystems and health (Zechendorf 1994). A major environmental concern is that herbicide resistance genes could be transferred to sexually compatible wild species (Dale 1992) resulting in wild relatives becoming herbicide-resistant weeds. From a health perspective there are concerns that marker-gene products may be allergenic or toxic and that the use of antibiotic resistance genes could, as a consequence of horizontal gene transfer, compromise the efficacy of antibiotics in clinical and veterinary applications.

Prior to commercialization of a transgenic crop a risk assessment evaluation is required, which includes assessment of the selectable marker gene and its product. Although there are no scientific or health and safety reasons to restrict the use of the *nptII* gene (Fuchs et al. 1993; FDA 1994) this may not be true of other selectable markers and a case-by-case examination of each will be necessary. Regardless of the assurances in risk-assessment reports, it is consumer acceptance that ultimately dictates the success of transgenic plants and produce; therefore, it would be prudent to alleviate perceived risks by elimi-

Molecular Methods of Plant Analysis, Vol. 22
Testing for Genetic Manipulation in Plants
Edited by J.F. Jackson, H.F. Linskens, and R.B. Inman
© Springer-Verlag Berlin Heidelberg 2002

nating selectable marker genes from transgenic crops prior to their field release and commercialization.

Eliminating the selectable marker gene from transgenic plants also has a number of practical benefits. Many of the new generation of transgenic crops contain not a single beneficial transgene but several different transgenes to make the crop of superior commercial value. For example, the development of transgenic plants engineered to produce novel metabolites or antibodies of therapeutic value has required the introduction of three or more transgenes (Ma et al. 1995; Slater et al. 1999). Crop improvement with multiple genes may require independent, sequential introduction of each gene into a cultivar to confer the new trait, and although multiple genes can be brought together through conventional genetic crossing, this is somewhat problematic for plants which are vegetatively propagated or have long generation times. Pyramiding genes by repeated transformation of an elite genotype could avoid the problems associated with crossing. However, the presence of a functional selectable marker gene precludes its use in subsequent retransformations and therefore different selectable markers would be required at each transformation step. Although a number of selectable marker genes are available for plant transformation (Yoder and Goldsbrough 1994), their use must be optimized empirically for the plant species of interest. Some marker genes have also proved to be of limited use in certain plant species due to a degree of tolerance to the selective agent even in the absence of the marker gene. The ability to eliminate a selectable marker gene after establishing a transgenic plant would allow the most optimal selection procedure to be used repeatedly in subsequent transformations. Several strategies have been developed to generate marker-free transgenic plants in a range of dicotyledonous and monocotyledonous plants and each is discussed below.

6.2 Co-transformation

The principle of the co-transformation strategy to generate marker-free transgenic plants is the integration of the transgene of interest and the marker gene into different unlinked locations of the plant genome and their subsequent segregation in the next generation to yield some progeny carrying the transgene but not the marker gene. *Agrobacterium tumefaciens*-mediated transformation is perhaps the most appropriate method of DNA delivery as it is well documented that many *Agrobacterium* strains harbor more than one T-DNA type and crown-gall tumors are often co-transformed with multiple T-DNAs (Hooykaas and Schilperoort 1992). The strategy requires a high efficiency of co-transformation and a high frequency of unlinked integration into the genome of the two T-DNA species.

Using a mixture of *A. tumefaciens* strains to deliver different T-DNAs, the co-transformation frequency of tobacco protoplasts was shown to be equal to the product of each single transformation event, indicating that co-transformation is the result of independent transformation events (Depicker et al. 1985). In the same study, the use of a single *A. tumefaciens* strain carrying two distinguishable T-DNAs on the same Ti plasmid resulted in higher co-transformation frequencies. This suggests that a single-strain approach may be the preferred option for achieving high-frequency co-transformation. However, De Block and Debrouwer (1991) observed that the frequency of co-transformation of hypocotyl explants of rapeseed (*Brassica napus*) cultivars using mixtures of *A. tumefaciens* strains to deliver different T-DNA molecules was higher than the product of two independent transformation events. This indicated that the co-transformation frequency may be significantly influenced by the choice of explant material. Despite the high co-transformation frequencies in rapeseed, the two distinct T-DNAs were mainly integrated at the same genomic locus. This linkage of the co-transformed T-DNAs was attributed to the use of nopaline-derived *A. tumefaciens* strains which predominantly result in inverted-repeat structures of integrated T-DNA (Jones et al. 1987; Jorgenson et al. 1987) or one of the two direct-repeat configurations (De Neve et al. 1997), irrespective of the plant species or transformation procedure. In addition, plants with linkage between different T-DNAs frequently have multiple copies of one form of the T-DNA in one or more of the four possible configurations. In contrast, a much less frequent occurrence of linked T-DNAs occurs with octopine-derived *A. tumefaciens* strains (Spielmann and Simpson 1986). McKnight et al. (1987) transformed tobacco leaf explants with a mixture of octopine-derived *A. tumefaciens* strains each carrying a different T-DNA on an otherwise identical binary vector, and although the co-transformation frequencies obtained were relatively low, analysis of the progeny of self-pollinated plants revealed that the different T-DNAs were unlinked. So although octopine-derived strains may result in lower co-transformation frequencies, the fact that the T-DNAs are often unlinked makes them more appropriate for the co-transformation strategy of producing marker-free transgenic plants. Komari et al. (1986) also used mixtures of octopine-derived *A. tumefaciens* strains, one harboring a binary vector with a marker gene (*npt*II or *hyg*) within the T-DNA and the second harboring a binary vector with the *gus* gene within the T-DNA. Variable co-transformation frequencies were observed in both tobacco and rice depending on the particular binary vectors used and the relative ratios of the mixture of strains. Komari et al. (1996) also developed binary vectors carrying two T-DNA molecules separated by a 15.2-kb spacer region. Using this vector in a single-strain approach resulted in significantly higher co-transformation frequencies in both rice and tobacco. Analysis of the progeny, after self-pollinating T_0 plants, revealed that many of the integrated T-DNAs were unlinked and selectable marker-free progeny that retained β-glucuronidase (GUS) expression could be recovered from more than half of the plants.

Another single-strain approach utilized *A. tumefaciens* harboring two binary vectors with compatible replicons (Daley et al. 1998). The T-DNA of the broad-host range pRK2-based vector included the *npt*II gene whilst that of the *A. rhizogenes* narrow-host range pRiHR1-based vector included the *gus* gene. Co-transformation frequencies of greater than 50% were observed for both rapeseed and tobacco and 40% of the co-transformed rapeseed plants when self-pollinated produced GUS$^+$ progeny which were sensitive to kanamycin, indicating that in these parental lines the co-transformed T-DNAs were unlinked. Self-pollination of the co-transformed tobacco plants revealed even higher levels of T-DNA segregation. PCR analysis of the rapeseed segregants revealed only one instance in which the *npt*II gene was present but not expressed, and although the presence of a non-functional marker was an uncommon event, it indicated the need for molecular analysis to provide definitive evidence of the plant being entirely free of the selectable marker gene.

As mentioned previously, T-DNA molecules often integrate into the genome as multiple copies in a direct- or inverted-repeat configuration. Although many of the studies described above demonstrated genetically that the different T-DNAs were unlinked and could be segregated in the progeny, the structural organization of the T-DNA was often not determined. One must consider the possibility that, in the segregated progeny, although only a single T-DNA type was present this may consist of T-DNA repeats. Putative marker-free plants must therefore be subjected to detailed molecular analysis as these T-DNA repeat structures are highly susceptible to gene silencing (Meyer and Saedler 1996; Depicker and Van Montegu 1997) and could subsequently result in the loss of transgene expression.

The co-transformation strategy has been used successfully to eliminate the selectable marker gene from both monocotyledonous and dicotyledonous transgenic plants produced via *Agrobacterium*-mediated transformation (Komari et al. 1996; Daley et al. 1998) and should be amenable to most, if not all, crop species which are sexually propagated. However, the degree of variability observed (see Table 6.1) depending on the *Agrobacterium* strain or strains, the use of a single or multiple binary vectors, the source of the explant material and its competence for transformation suggests that maximizing the efficiency of the co-transformation strategy will require optimization of a number of transformation parameters for the plant species of interest.

DNA delivery via particle bombardment is an alternative method of plant transformation and has been used to co-transform DNA molecules. Co-transformation frequencies in excess of 70% have been reported after bombardment of plant tissue with a marker gene and a transgene, located on separate plasmids (Spencer et al. 1992; Wakita et al. 1998). However, the transgene and marker gene invariably fail to segregate in the progeny of self-pollinated T$_0$ plants, as both DNA molecules are integrated as multiple copies into the same loci of the genome. This high incidence of linkage means that the co-transformation strategy is unlikely to be of use in eliminating marker genes from transgenic plants produced via particle bombardment transformation.

Table 6.1. Co-transformation and T-DNA segregation of transgenic plants produced via *Agrobacterium*-mediated transformation. *nr* Not reported, *nd* not determined

A. tumefaciens strain type	vir Region	Single or mixed	Plant species (explant)	Co-transformation frequency	Co-transformants with segregating T-DNAs	Reference
C58C1Rif^R	Octopine/Nopaline	Mixed	Tobacco (protoplasts)	45% (68/150)	nd	Depicker et al. (1985)
C58C1Rif^R	Nopaline	Single[a]	Tobacco (protoplasts)	67% (64/96)	nd	Depicker et al. (1985)
LBA4404	Octopine	Mixed	Tobacco (leaf disc)	19% (3/16)	100% (3/3)	McKnight et al. (1987)
C58C1Rif^R	Nopaline	Mixed	Rapeseed "Drakkar" (hypocotyl)	39%	33% (3/9)	De Block and Bebrouwer (1991)
C58C1Rif^R	Nopaline	Mixed	Rapeseed "Westar" (hypocotyl)	85%	nd	De Block and Bebrouwer (1991)
C58C1Rif^R	Nopaline	Mixed	Rapeseed "R8494" (hypocotyl)	56%	nd	De Block and Bebrouwer (1991)
LBA4404	Octopine	Single[a]	Tobacco (leaf disc)	51% (115/227)	79% (15/19)	Komari et al. (1996)
LBA4404	Octopine	Mixed	Tobacco (leaf disc)	18% (57/310)	71% (10/14)	Komari et al. (1996)
LBA4404	Octopine	Single[a]	Rice (scutellar)	47% (259/549)	65% (13/20)	Komari et al. (1996)
LBA4404	Octopine	Mixed	Rice (scutellar)	2% (2/82)	100% (2/2)	Komari et al. (1996)
LBA4404	Octopine	Mixed[c]	Rice (scutellar)	14% (7/49)	nd	Komari et al. (1996)
C58C1Rif^R	Octopine	Mixed	Tobacco (protoplasts)	nr	71% (10/14)	De Neve et al. (1997)
C58C1Rif^R	Octopine	Mixed	Arabidopsis (leaf disc)	nr	42% (5/12)	De Neve et al. (1997)
C58C1Rif^R	Octopine	Mixed	Rapeseed (hypocotyl)	nr	40% (4/10)	De Neve et al. (1997)
LBA4404	Octopine	Single[b]	Arabidopsis (root)	62% (21/34)	40% (8/20)	Daley et al. (1998)
LBA4404	Octopine	Single[b]	Tobacco (leaf disc)	52% (52/100)	59% (24/41)	Daley et al. (1998)

[a] T-DNAs on different replicons.
[b] Two T-DNAs on the same replicon.
[c] 3:1 mix of *Agrobacterium* strains.

6.3 Transposon-Mediated Approaches

Two transposon-mediated strategies have been developed to generate marker-free transgenic plants. These strategies involve *Agrobacterium*-mediated transformation followed by intra-genomic relocation of the transgene of interest, and its subsequent segregation from the selectable marker in the progeny (Goldsbrough et al. 1993) or excision of the marker gene from the genome (Ebinuma et al. 1997). Both strategies were developed using the maize *Ac/Ds* transposable element but could be adapted to use similar autonomous transposable elements.

The features of the autonomous *Ac* element essential for its transposition are the 5′- and 3′- terminal repeat sequences and the transposase function encoded by ORFa (reviewed in Fedoroff 1989). *Ds* elements lack a transposase and are stable within the genome but can be transactivated by the *Ac* element, as can other sequences flanked by the terminal repeats. Transposition results in excision of the *Ac* or *Ds* element from the genome and its reinsertion into another locus, and although there is a preference for transposition to a linked locus (Jones et al. 1990) it is not uncommon for transposition to chromosomal locations of sufficient genetic distance to allow recombination between the primary and secondary sites (Belzile et al. 1989). Moreover, *Ac/Ds* elements maintain their transposition functions in other plant species, facilitating their use as a genetic tool in these plants.

6.3.1 Transposon-Mediated Repositioning

Goldsbrough et al. (1993) developed transformation vectors incorporating *Ac/Ds* elements to facilitate the physical separation of the transgene of interest from the selectable marker by transposition to a secondary chromosomal location after transformation (Fig. 6.1A). If transposon-mediated relocation is of sufficient genetic distance from the primary integration locus then self-pollinated or out-crossed plants can produce progeny lacking the marker gene but retaining the transgene. The T-DNA used contained an *npt*II marker, *Ac* transposase and a chimeric *Ds/gus* element, in which *gus* was flanked by the 5′- and 3′-terminal repeat sequences of *Ac*. After self-pollination of a tomato transformant with a single T-DNA integration approximately half of the GUS$^+$ progeny were shown to have at least one *Ds/gus* element at a locus different to that of the primary integration locus. In two of these GUS$^+$ progeny the absence of the *npt*II and transposase genes indicated that segregation had occurred as a consequence of the relocation of the *Ds/gus* element to an unlinked locus. Upon self-pollination of a primary transformant with two T-DNA insertions, progeny were identified with the *Ds/gus* element present in the absence of all other original T-DNA elements at a frequency of 7%. This frequency was in fact higher than the occurrence of such progeny from the single T-DNA trans-

A

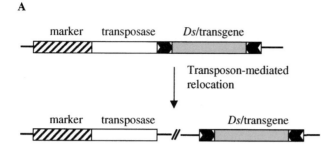

B

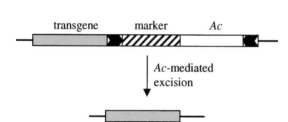

Fig. 6.1A,B. Transposon-mediated strategies to marker-gene removal. **A** Transposon-mediated relocation strategy. **B** Transposon-mediated excision strategy. The marker gene is represented by the *hatched boxes*, transgene by the *shaded boxes*, transposase by the *unshaded boxes* and inverted terminal repeats by the *black arrows*

formant and was probably a consequence of higher levels of transposase activity in the transformant with two T-DNA inserts. This study demonstrated that, through the occurrence of germinally inheritable transposon-mediated relocation, progeny can be isolated that are free of marker genes but retain the transgene, even when the primary transformant contains multiple T-DNA insertions. There is, however, a note of caution in adopting the strategy as described by Goldsbrough et al. (1993) to pyramid multiple transgenes. Repeated use of this transposition strategy to remove the marker gene of a second transformation event could result in the transposition of all *Ds*/transgene elements. This would be problematic if a particular transgenic line had been selected on the basis of desirable expression of the first transgene, as transposition to a new chromosomal location may significantly alter its expression. Therefore, it would be more prudent to flank the transposase and marker gene with *Ac* terminal repeats and relocate these genes, rather than the transgene of interest.

6.3.2 Transposon-Mediated Elimination

The transposon-mediated strategy described above is reliant on the crossing of plants to segregate the marker and transposase from the relocated transgene. As a consequence, this strategy is of limited use for vegetatively propagated plants or plants with long reproductive cycles. An adaptation to the relocation strategy, avoiding the need for crossing, is the transposon-mediated excision strategy. This strategy exploits the fact that in approximately 10% of *Ac* transposition events the *Ac* element fails to reintegrate into the genome or reintegrates into a sister chromatid which is subsequently lost by somatic segregation (Belzile et al. 1989). In this strategy the T-DNA is comprised of a chimeric *Ac* element which includes the selectable marker and transposase genes, whereas the transgene of interest, which will be retained in its original genomic location, resides outside the chimeric *Ac* element (Fig. 6.1B).

Ebinuma et al. (1997) demonstrated the feasibility of this strategy by eliminating the isopentenyl transferase (*ipt*) marker gene from transgenic tobacco plants. Transgenic plants constitutively expressing the *ipt* gene have elevated cytokinin to auxin ratios resulting in a loss of apical dominance, suppression of root formation and what is referred to as a "shooty" phenotype (Smigocki and Owens 1989). When Ebinuma et al. (1997) transformed tobacco leaf-discs with a T-DNA containing *npt*II and *gus* genes and a chimeric *Ac* element which included a 35S-*ipt* gene, two thirds of the differentiated adventitious shoots showed an extreme shooty phenotype. Upon culturing these phenotypically distinct shoots morphologically normal shoots developed, indicating loss of the *ipt* gene expression, at a frequency of 0.032%. Analysis of these normal shoots revealed that the *Ac/ipt* element had indeed been excised from the genome. The shoots retained GUS and NPTII activity indicating that excision had been limited to the *Ac/ipt* element and that the rest of the T-DNA remained integrated in the genome. Although this demonstrated that transposon-mediated excision has the potential to be applied to eliminating selectable marker genes from transgenic plants, the frequency observed was lower than expected. Given that the frequency of *Ac* transposition events that occur in tobacco plants and are transmitted to the progeny are 1–5% (Jones et al. 1990) and that *Ac* elements are eliminated from the genome during 10% of these events, then the frequency of somatic elimination of *Ac* from the shoot apical meristem should be 0.1–0.5%. Ebinuma et al. (1997) established that the lower than expected frequency they observed was due to the primary transformants containing multiple T-DNA copies and, as a consequence, a normal shoot phenotype would only occur if all the copies of the *ipt* gene had been eliminated from the genome. Therefore, minimizing the frequency of multiple T-DNA insertions would probably increase the frequency of obtaining marker-free transgenic plants through this transposon-mediated elimination strategy.

Owing to the fact that the *npt*II gene was retained in the T-DNA, no truly marker-free plants were generated by Ebinuma et al. (1997). However the *npt*II gene was used merely as a reporter gene and not in the selection of either the normal or abnormal shoot phenotypes. The utilization of the *ipt* gene provided a relatively simple visual marker, firstly for transformants carrying the T-DNA and secondly to select transposon-mediated elimination events. As constitutive *ipt* gene expression results in similar phenotypes in other plant species its use as a visual marker in a transposon-mediated marker excision strategy should be applicable to at least these, and perhaps other, plant species.

A significant advantage of this strategy over those previously described is that marker-free transgenic plants can be selected at the T_0 generation, avoiding the need to sexually cross plants and thereby making the strategy applicable to vegetatively propagated crops (e.g. potato, grape, apple) and plants with long reproductive cycles (e.g. forest trees). Use of this strategy has been reported in hybrid aspen in which the appearance of normal shoots from shoots with an extreme shooty phenotype was a consequence of excision of the *Ac/ipt* element (Ebinuma et al. 1997).

6.4 Site-Specific Recombination

A number of prokaryotes and lower eukaryotes encode site-specific recombination systems that utilize one or more proteins which act to cause defined recombination between specific DNA target sequences. Site-specific recombinases of the integrase and invertase families have been used to manipulate DNA in heterologous cellular environments (reviewed in Kilby et al. 1993; Ow and Medberry 1995). Three integrase systems have been shown to function in dicotyledonous and monocotyledonous plants: the Cre/*lox*P of *Escherichia coli* bacteriophage P1; the Flp/*frt* of the *Saccharomyces cereviseae* 2-μm plasmid; and the *R*/RS of the *Zygosaccharomyces rouxii* pSR1 plasmid (reviewed in Odell and Russell 1994). In addition, a modified Gin/*gix* invertase system of bacteriophage Mu has been shown to function in dicotyledonous plant cells (Maeser and Kahmann 1991). All are simple two-component systems requiring a single protein, which can act *in trans,* to cause DNA cleavage at a precise position in specific asymmetric target sequences, followed by ligation of two cleaved target sites. If both target sites are *in cis* this results in inversion or deletion of the intervening DNA depending on the relative orientation of the target sites. It is the asymmetry of the target sequence that gives them their directionality. Deletion of intervening DNA between two target sites in the same orientation, coupled with the fact that no additional host factors are required, offers the opportunity to utilize these recombination systems in eliminating the selectable marker gene from transgenic plants. The basic principle of such a strategy is to generate transgenic plants in which the marker gene is flanked by recombinase target sequences in the same orientation, and is excised from the genome upon introduction and expression of the recombinase (Fig. 6.2A).

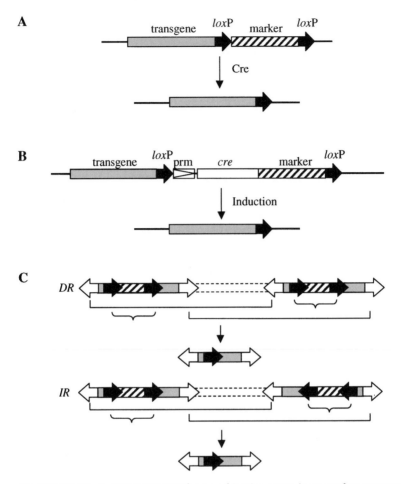

Fig. 6.2A–C. Cre-mediated site-specific recombination strategies to marker-gene removal. **A** The basic site-specific recombination strategy. **B** The "single step" excision strategy. **C** Marker-gene excision and resolution of multiple copies of DNA integrated into the genome in an inverted repeat (*IR*) or direct repeat (*DR*) structure. The marker gene is represented by the *hatched boxes*, the transgene by the *shaded boxes*, *lox*P sites by the *black arrows* and variant *lox*P sites by the *white arrows*

6.4.1 The Cre/*lox*P System

The 38-kDa Cre recombinase recognizes a 34-bp *lox*P site consisting of two 13-bp inverted repeats flanking an 8-bp core. The normal function of Cre is to resolve dimers of the plasmid form of bacteriophage P1 through recombination at *lox*P sites. Cre-mediated recombination at *lox*P sites was first demonstrated in plants by Dale and Ow (1990) and Odell et al. (1990). Dale and Ow (1991) subsequently reported the use of the Cre/*lox*P system to eliminate the marker gene from transgenic tobacco plants containing a T-DNA which included a 35S-luciferase gene and a 35S-*hyg* gene flanked by *lox*P sites, in the same orientation. Retransformation of Hyg^R Luc^+ plants with a 35S-*cre* and *npt*II T-DNA resulted in 91% of the kanamycin-resistant plants being susceptible to hygromycin due to precise Cre-mediated recombination between the *lox*P sites and excision of the intervening *hyg* gene from the genome. In addition, there was no evidence that the excised DNA had reintegrated at another genomic location. Providing the T-DNA introduced in the retransformation event integrated into a locus unlinked to the original T-DNA, then, upon self-pollinating, up to 25% of the progeny showed luciferase expression and the absence of both the *npt*II and *hyg* marker genes. Similarly, retransformation of transgenic tobacco plants containing a 35S-*gus* and *lox*P-flanked 35S-herbicide-resistant acetolactate synthase (ALS^R) gene with a 35S-*cre* and *hyg* T-DNA resulted in 95% of the Hyg^R plants showing no chlorsulfuron resistance as a consequence of recombination at *lox*P sites and ALS^R gene excision (Russell et al. 1992). In addition, almost all progeny from self-pollinated retransformants failed to germinate in the presence of the herbicide. These high frequencies of excision of the *lox*P-flanked marker and the fact that the plants showed no indication of being chimeric at the *lox*P locus suggest that this retransformation approach to recombinase delivery results in efficient Cre-mediated excision, which occurs early (between T-DNA introduction and plant organogenesis), and that the recombination event is germinally inherited.

Dale and Ow (1991) also crossed their *luc*/*hyg* and *cre*/*npt*II plants revealing that approximately half of the Kan^R Luc^+ progeny were sensitive to hygromycin. This alternative approach to recombinase delivery also resulted in Cre-mediated excision of the *lox*P-flanked *hyg* gene, although at a considerably lower efficiency than that achieved through retransformation. Russell et al. (1992) observed that GUS^+ progeny of crosses between homozygous *cre*/*hyg* and heterozygous single T-DNA locus ALS^R/*gus* plants showed a variation in chlorsulfuron sensitivity. These degrees of herbicide sensitivity were a consequence of many progeny being chimeric for the presence of the *lox*P-flanked ALS^R gene, with the ratios of excision to non-excision varying between different crosses. Also, many seedlings exhibited herbicide damage only in the newly formed leaves indicating that somatic ALS^R excision was occurring in the developing seedling. In essentially identical experiments using *Arabidopsis* all Hyg^R GUS^+ F1 progeny were sensitive to chlorsulfuron in a callus-induction assay using internodal stem tissue. However, after allowing these plants to produce new inflorescences and self-pollinate, some plants gave rise to chlorsulfuron-resistant F2 seed or seed of an intermediate resistant phenotype. Therefore, although chimerism was not apparent in the primary stem it had become apparent in the F2 seedlings.

These findings demonstrated that introduction of the recombinase via cross-pollination can sometimes result in early and efficient Cre-mediated excision of the *lox*P-flanked marker gene. However, unlike the retransformation approach, crossing results in many progeny retaining the *lox*P-flanked marker gene and a high incidence of progeny which are chimeric at the *lox*P locus. The differences in efficiency between the two approaches to recombinase delivery may be a consequence of the different cell types in which the recombination reaction must take place to produce a plant homogeneous for marker gene excision and/or differences in recombinase expression in these cell types. As retransformation consistently resulted in high-efficiency recombination and plants homogeneous for excision it appears that the *cre* gene is sufficiently expressed from the 35S promoter in recombination-receptive progenitor cell types early in the development of shoots from leaf discs. However, the frequent heterogeneity of excision observed in the cross-pollination approach suggests either insufficient *cre* expression from the 35S promoter in the embryo or apical meristem, or that the cells of these tissues are less receptive to recombination. Russell et al. (1992) observed that, in crossing plants, excision efficiency was influenced by the level of recombinase expression in the parental plant. Their findings revealed that the parent with the lowest *cre* mRNA level in leaf tissue was the most efficient in promoting excision. However, one must interpret this finding with caution as *cre* mRNA levels in leaves may differ considerably from those in developing embryos and germinating seedlings. Irrespective of the recombinase delivery approach used, the efficiency of excision does not appear to be influenced by the *lox*P T-DNA integration site, or in the cross-pollination approach by whether the male or female parent carries the recombinase.

Generating marker-free transgenic plants via these Cre-*lox*P strategies requires at least one instance of sexual crossing. This is due to both approaches resulting in the introduction of a *cre*-linked selectable marker gene which must be segregated away from the transgene of interest. This need to cross plants is therefore a limitation in generating marker-free vegetatively propagated plants or plants with long reproductive cycles through the use of these Cre/*lox*P strategies.

An adaptation of the retransformation strategy which obviates the need for segregation of the *cre*-linked marker from the transgene of interest was demonstrated by Gleave et al. (1999). This approach used transient *cre* expression, exploiting the observations that during *Agrobacterium*-mediated transformation the number of plant cells to which T-DNA is transferred greatly exceeds the number of cells which become stably transformed, and that genes are expressed from non-integrated T-DNA (Janssen and Gardner 1989). The strategy also used the cytosine deaminase (*cod*A) gene, which in the presence of 5-fluorocytosine (5-Fc) acts as a negative selectable marker (Stougaard 1993), to select for concomitant Cre-mediated excision of the *npt*II and *cod*A genes. Tissue from tobacco plants with a single T-DNA insert of *lox*P-flanked *npt*II and 35S-*cod*A genes and a 35S-*gus* outside the *lox*P-flanked region was cocultivated with *A. tumefaciens* harboring a 35S-*cre* and 35S-*hyg* T-DNA. Less than 3% of the 773 regenerated shoots survived under 5-Fc selection. Two of these plants had regenerated from cells which had undergone Cre-mediated *npt*II and *cod*A excision as a consequence of transient *cre* expression from T-

DNA that had failed to integrate into the genome. Through this transient recombinase expression strategy, marker-free transgenic plants were produced without any requirement for sexual crossing, although the frequency (<0.3%) with which they were recovered suggests that the ability to select for recombinase-mediated excision events is essential if the strategy is to be of any practical use.

Another Cre/loxP adaptation that avoids the need to cross plants is referred to as "single-step" excision (Surin et al. 1997). Here the cre gene is included with the selectable marker gene within the loxP-flanked region of the T-DNA, and the transgene of interest resides outside the loxP-flanked region (Fig. 6.2B). This strategy relies on suppression of recombinase expression during the initial stages of transformation and release of the suppression after establishment of a transgenic plant to allow recombination of the loxP-flanked regions of the T-DNA and excision of the marker gene and the cre gene. To demonstrate this, Surin et al. (1997) placed the cre gene under the transcriptional control of the light-regulated Arabidopsis rbcS 1a promoter. The T-DNA of the binary vector was organized such that excision of the loxP-flanked rbcS 1a-cre and nptII genes would lead to activation of GUS expression by the subterranean clover stunt virus Sc4 promoter. Initially, cre expression was suppressed by regeneration of transformed tissue in the dark and then plants were placed under normal light conditions and the kanamycin selection pressure removed. GUS expression in the vascular pattern expected for the Sc4-GUS was observed in leaves of 28% of the regenerated plants, although all of these plants were found to be chimeric with respect to the loxP-flanked region of the T-DNA. Subsequent regeneration of plants from leaves of these chimeric T_0 plants resulted in 6–92% of the regenerants being homogeneous for excision of the loxP-flanked region, with the variation being dependent upon the T_0 parent. It appears that successful application of this strategy will be very much reliant on the tight control of recombinase expression and could perhaps make use of one of the numerous chemically induced promoters that are now available for plant gene expression (Gatz and Lenk 1998).

All of the Cre/loxP strategies discussed so far have the potential to be exploited to generate marker-free transgenic plants in a range of crop species. However, one consideration is that, upon excision of the loxP-flanked region, one copy of the loxP sequence remains in the genome. Therefore, repetitive use of the Cre/loxP system in subsequent marker-gene elimination could lead to genome deletions, inversions or translocations due to recombination between loxP sites of a secondary T-DNA and the loxP site remaining from the initial excision event in the primary T-DNA. This could be overcome through the use of identical pairs of loxP sequence variants at each transformation step which fail to recombine with dissimilar loxP sites as a consequence of base changes in the asymmetric core (Hoess et al. 1986).

The marker-gene elimination strategies discussed above have all been developed in transgenic plants produced via Agrobacterium-mediated transformation, and their utility is simplified by the presence of single, intact T-DNA molecules integrated into the genome. In certain plant species, particularly cereal crops (Kohli et al. 1998), the incidence of a single, intact copy of integrated DNA is uncommon, irrespective of whether transformation is via *Agrobacterium* or particle bombardment. Multiple copies of integrated DNA occur in both tandem direct and inverted repeats and these structures are highly susceptible to gene silencing (Meyer and Saedler 1996) and not conducive to the marker-gene elimination strategies discussed so far.

Srivastava et al. (1999) developed a Cre/*lox*P strategy to not only eliminate the marker gene but to also resolve multiple transgene copies to a single copy. This strategy uses two pairs of recombination target sites within the integrating DNA, one pair being *lox*P sites and the second pair being a variant (e.g. *lox*511) which fails to recombine with the *lox*P sites (Hoess et al. 1986). The marker gene is flanked by *lox*P sites in the same orientation, whilst the entire integrated DNA is flanked by *lox*511 sites in an inverted orientation. This organizational arrangement means that, at a single locus, irrespective of their orientation, multiple copies of the integrated DNA are flanked by *lox*511 sites in an inverted orientation (Fig. 6.2C). If one imagines a single locus of multiple DNAs from left to right, then recombination of the first *lox*511 with the penultimate *lox*511 or recombination between the second *lox*511 and the last *lox*511 would resolve the multiple copy locus into a single integrated DNA molecule. Additional recombination between *lox*P sites would excise the marker gene. Srivastava et al. (1999) produced transgenic wheat plants via particle bombardment and crossed plants expressing the *bar* selectable marker gene and a single locus of inverted and/or direct repeats of the integrated DNA with transgenic plants expressing the *cre* gene. Analysis of the F1 progeny revealed that the *lox*P-flanked *bar* gene had been excised and only a single copy of the integrated DNA remained in the genome. By self-pollination of these F1 plants the *cre* locus and its linked marker gene could be segregated from the transgene locus to produce transgenic wheat devoid of marker genes. However, 20–40% of F2 progeny, derived from F1 plants which appeared to be resolved at the integration locus, showed evidence of multiple transgene copies suggesting that germline cells in the F1 had been chimeric. Nevertheless, single-copy marker-free transgenic plants were produced relatively efficiently from plants with a single integration locus despite this locus originally containing multiple copies of the integrated DNA molecule.

6.4.2 The FLP/*frt* System

The FLP/*frt* site-specific recombinase of the 2-µm plasmid of *Saccharomyces cerevisiae* is analogous to the Cre/*lox*P recombinase. The 48-kDa FLP recombinase acts at *frt* sites and can promote excision of intervening DNA between two *frt* sites in the same orientation. The 48-bp *frt* sequences consist of three 13-bp imperfect repeats, two in an inverted orientation flanking an 8-bp asymmetric core and the third, which can be removed without any loss of recombination efficiency, in the same orientation as the repeat located on the same side of the core. The FLP/*frt* recombinase system has been shown to function in both monocotyledonous and dicotyledonous plant cells (Lyznik et al. 1993; Lloyd and Davis 1994).

To demonstrate the potential of the FLP/*frt* system in eliminating selectable marker genes from plant cells, Lyznik et al. (1995) produced a recombination target vector with the maize *Ubi-1* promoter transcribing *npt*II, followed by a promoterless *gus* gene. Target *frt* sequences, in the same orientation, flanked the *npt*II gene, with the second of these (*frt*m) lacking the third repeat element of the *frt* sequence. FLP-mediated recombination between *frt* and *frt*m would therefore result in *npt*II excision and activation of *gus* expression by the *Ubi-1* promoter. Heat treatment of maize protoplasts co-transformed with the target vector and a vector carrying FLP transcribed from a soybean heat-shock promoter resulted in 20–25% of the target vector undergoing *npt*II excision. A decrease in NPTII activity and concomitant GUS expression was also observed after heat treatment in 40% of callus lines stably transformed with both vectors. GUS activity was also observed in one-third of the hygromycin-resistant calli of protoplasts, derived from a maize callus line stably transformed with a single copy of the target vector, after co-transformation with 35S-*hyg* and *Ubi 1*-FLP vectors (Lyznik et al. 1996). However, although GUS activity was due to precise FLP-mediated recombination between the *frt* and *frt*m, the majority of calli also showed evidence of additional rearrangements, reintegration and chimerism.

Crossing tobacco plants transformed with a *frt*-flanked *hyg* gene, located between a 35S promoter and a *gus* gene, with 35S-FLP transgenic plants resulted in FLP-mediated excision of the *frt*-flanked *hyg* gene and GUS$^+$ progeny (Kilby et al. 1995). However, as with the Cre/*lox*P strategy in which the target and recombinase are brought together through crossing, chimeric progeny were also identified. Kilby et al. (1995) also crossed *Arabidopsis* transformed with the *frt*-target cassette with *Arabidopsis* expressing FLP transcribed from a heat-shock promoter. Heat treatment resulted in approximately 10% of progeny of these crosses showing clonal sectors of GUS expression, due to excision of the *frt*-flanked *hyg* gene.

Although not evaluated to the same extent as the Cre/*lox*P system in developing marker-gene elimination strategies, these studies provide sufficient evidence to suggest that the FLP/*frt* system could be used in most, if not all, of the Cre/*lox*P strategies described in Sect. 6.4.1. It is also likely that Flp/*frt* strategies would be subject to the same limitations as the Cre/*lox*P strategies. Another limitation of the FLP/*frt* system is that FLP expression may be detrimental to some plant species. For example, there are reports of unsuccessful attempts to generate transgenic *Arabidopsis* constitutively expressing FLP (Lloyd and Davis 1994; Kilby et al. 1995). Detrimental effects have also been observed in plants constitutively expressing the modified Gin recombinase (Maeser and Kahmann 1991) probably due to the occurrence of endogenous sequences in the plant genome recognized by the recombinase.

6.4.3 The *R/RS* System

The *R/RS* system of the *Zygosaccharomyces rouxi* pSR1 plasmid comprises the
56-kDa R recombinase which acts at 58-bp RS sequences with two 12-bp
inverted repeats either side of a 7-bp asymmetric core. Onouchi et al. (1991)
established that the *R/RS* system was functional in tobacco cells in both R-
mediated excision and inversion, depending on the relative orientation of the
RS sites. This was followed by a study in which transgenic *Arabidopsis* homozygous for a 35S-R gene were crossed with plants homozygous for an RS-flanked
35S-CAT cassette, which when excised would result in *gus* activation (Onouchi
et al. 1995). The sectorial chimerism of GUS expression in the F1 progeny was
an indication that R-mediated excision occurred early in leaf development or
in discrete groups of cells in the immature shoot apical meristem. After self-
pollination, GUS expression was observed in all tissues of 0.1–2.4% of F2
progeny, indicating that they had arisen from germline cells that had undergone R-mediated excision. These studies show that the *R/RS* system is
amenable to many of the recombinase strategies described earlier and is also
likely to be subject to the same limitations.

In Sect. 6.3.2, a transposon-mediated excision strategy to generating
marker-free transgenic plants without the need for sexual crossing was
described (Ebinuma et al. 1997). This strategy was subsequently adapted to
utilize the *R/RS* system (Sugita et al. 1999). The T-DNA used consisted of a 35S-
R and 35S-*ipt* cassette flanked by RS sequences and 35S-*gus* and *npt*II genes
outside of the RS-flanked region. Three types of adventitious buds differentiated from tobacco leaf-discs transformed with this T-DNA on hormone-
free media: phenotypically normal, moderate shooty, and extreme shooty.
Reporter-gene expression and PCR analysis showed that R-mediated excision
of the RS-flanked R and *ipt* genes had occurred even prior to formation of the
shooty phenotype in 13% of the phenotypically normal shoots, and over time
normal shoots appeared from a high proportion of the moderate and extreme
shooty lines. R-mediated excision occurred in a third of the normal shoots
derived from moderate shooty lines and two thirds of those derived from the
extreme shooty lines, many of which contained multiple T-DNA copies. Despite
this presence of multiple T-DNAs, and therefore multiple RS sites, no phenotypes attributable to aberrant recombination events were observed, although
confirmation that R-mediated genomic deletions, inversions or translocations
had not occurred would require detailed molecular analysis. This study showed
that marker-free transgenic plants retaining expression of the transgene(s)
could be recovered from *ipt*-shooty phenotypes at a frequency eight times
higher than had been observed in the transposon-mediated excision strategy.
Sugita et al. (2000) further improved this *R/RS* strategy by expressing the *ipt*
gene from its native promoter and the *R* gene from the safener-inducible promoter of the maize glutathione-S-transferase gene. After transformation of
tobacco with the modified construct the *ipt*-shooty phenotype developed from
almost 90% of the adventitious buds. Subsequent safener induction resulted in

25% of these *ipt*-shooty lines giving rise to phenotypically normal, GUS[+] shoots due to R-mediated excision of *ipt*, and the majority of these marker-free transgenic plants contained a single copy of the transgene.

6.5 Intrachromosomal Homologous Recombination

The exploitation of intrachromosomal homologous recombination in plants to produce genomic deletions has been limited by the very low frequencies with which such events occur (Puchta et al. 1995). Zubko et al. (2000) suggested that these frequencies could be increased by using sequences which provide a more efficient substrate for the plant's recombination machinery, and they selected the *att*P sequences of phage λ to evaluate this hypothesis. Integration of phage λ into its host genome occurs by recombination between the *att*P region and the *E. coli att*B region and requires the phage-encoded integrase (Int) and the bacterial-encoded integration host factor (IHF). Zubko et al. (2000) produced a T-DNA that comprised a transformation booster sequence, which enhances homologous and illegitimate recombination (Galliano et al. 1995), an oryza-cystatin-I gene, and an *npt*II and *tms*2 cassette flanked by 352 bp of *att*P sequences. When shoots were regenerated from kanamycin-resistant tobacco calli, after a period during which kanamycin selection had been temporarily removed, two of eleven clones produced a mixture of green and white shoots. The appearance of the white shoots was indicative of excision of the *npt*II gene. Approximately half of the shoots regenerated from these white leaves were capable of rooting in the presence of naphthalene acetamide (NAM). This was indicative of *tms*2 excision, as *tms*2 expression results in the conversion of NAM to the auxin NAA which suppresses root development. Although PCR analysis confirmed the absence of the *npt*II and *tms*2 genes, only three of 23 plants retained the oryzacystatin-I transgene, indicating that excision had occurred beyond the *att*P sequences. Two of these plants were homogeneous for the recombination event and a single *att*P sequence remained in the genome, suggesting precise intrachromosomal recombination between the *att*P sequences, even though the Int and IHF proteins had not been expressed. These findings indicate that the intrachromosomal recombination was not exclusively associated with homologous recombination between *att*P sequences and that illegitimate recombination remained the dominant form of intrachromosomal recombination, accounting for the loss of the transgenes outside the *att*P-flanked region.

Although it is not possible to determine the actual intrachromosomal recombination frequency from the data presented by Zubko et al. (2000), it seems that such recombination events would only be detected if they occurred at a much higher frequency than has previously been reported or if they occurred very early in shoot development to give rise to large clonal sectors. Higher frequencies of recombination than previously reported may be due to the presence of the *att*P sequences, which through their natural function may have a secondary structure conducive to recombination. However, the possibility cannot be excluded that higher recombination frequencies are a consequence of the use of undifferentiated callus rather than differentiated plant cells, the presence of the transformation booster sequence or the genomic location of the integration site. Therefore it remains to be determined whether the *att*P sequences do indeed increase the frequency of intrachromosomal homologous recombination, although this study does highlight the possibility of using sequence elements that may be highly recombinogenic to develop new strategies for generating marker-free transgenic plants.

6.6 Conclusions and Future Prospects

Although selectable marker genes have been a necessity in plant transformation, there are now good scientific, environmental and commercial reasons for their elimination from transgenic plants. Theoretically it may be possible to avoid the use of selectable markers and employ high-throughput automated PCR approaches to identify transgenic plantlets or calli, although the practicalities of such an approach will be highly demanding. Therefore, while the frequencies of DNA integration into the plant genome remain low, selectable markers will remain a necessity in the efficient production of transgenic plants. This being the case, there is often a need to eliminate these genes once a transgenic plant has been established, and a range of strategies are now available and have been successfully exploited in laboratory studies to achieve this goal. All of these strategies, however, have their limitations, and to the author's knowledge none has yet been utilized in the production of transgenic crops for commercial use. The strategy that one adopts to produce marker-free transgenic plants will very much depend on the plant species and the method of transformation used. Perhaps the simplest strategy is that of co-transformation, in that it may require very little alteration to existing transformation vectors and procedures and could be used repeatedly to pyramid multiple genes. Its major limitation is probably the reliance on sexual crossing and therefore its inability to be used for vegetatively propagated crops. Of the more elaborate approaches, site-specific recombination strategies have proven successful in a range of dicotyledonous and monocotyledonous plants and can result in high frequencies of homogeneous marker-free transgenic plants. Site-specific recombination and transposon-mediated strategies have also been adapted or purposely designed to be applicable to vegetatively propagated crops, although the efficiencies of some of these approaches would need improvement for them to be of practical use. Other site-specific recombination strategies have been adapted to deal with complex integration loci of multiple DNA molecules, which are all too common in transgenic plants of commercial significance. Applying the more elaborate strategies to crop plants will probably require considerable adaptation of existing transformation vectors to introduce the necessary elements required for recombination or transposition-mediated excision. Further alterations to vectors may also be required if the strategy is to be used repeatedly to pyramid multiple transgenes in a transgenic cultivar. With a few exceptions, demonstration of feasibility and success of these strategies has been carried out largely in model plant species, and it remains to be seen whether similar frequencies of recovering marker-free transgenic plants can be achieved with plants of commercial value. Nevertheless, the use of the range of options that now exists for producing marker-free transgenic plants should be considered in plant transformation programs, particularly those which are heading towards commercialization.

References

Belzile F, Lassner MW, Tong Y, Khush R, Yoder JI (1989) Sexual transmission of transposed activator elements in transgenic tomatoes. Genetics 123:181–189

Dale EC, Ow DW (1990) Intra- and intermolecular site-specific recombination in plant cells mediated by bacteriophage P1 recombinase. Gene 91:79–85

Dale EC, Ow DW (1991) Gene transfer with subsequent removal of the selection gene from the host genome. Proc Natl Acad Sci USA 88:10558–10562

Dale PJ (1992) Spread of engineered genes to wild relatives. Plant Physiol 100:13–15

Daley M, Knauf VC, Summerfelt KR, Turner JC (1998) Co-transformation with one *Agrobacterium tumefaciens* strain containing two binary plasmids as a method for producing marker-free transgenic plants. Plant Cell Rep 17:489–496

De Block M, Debrouwer D (1991) Two T-DNAs co-transformed into *Brassica napus* by a double *Agrobacterium tumefaciens* infection are mainly integrated at the same locus. Theor Appl Genet 82:257–263

De Neve M, De Buck S, Jacobs M, Van Montegu M, Depicker A (1997) T-DNA integration patterns in co-transformed plant cells suggest that T-DNA repeats originate from co-integration of separate T-DNAs. Plant J 11:15–29

Depicker A, Van Montegu M (1997) Post-transcriptional gene silencing in plants. Curr Opin Cell Biol 9:373–382

Depicker A, Herman L, Jacobs A, Schell J, Van Montagu M (1985) Frequencies of simultaneous transformation with different T-DNAs and their relevance to *Agrobacterium*/plant cell interactions. Mol Gen Genet 20:477–484

Ebinuma H, Sugita K, Matsunaga E, Yamakado M (1997) Selection of marker-free transgenic plants using the isopentenyl transferase gene. Proc Natl Acad Sci USA 94:2117–2121

FDA (1994) Secondary direct food additives permitted in food for human consumption: food additives permitted in feed and drinking water of animals: aminoglycoside-3'-phosphotransferase II: final rule. Fed Reg 59:26700–26711

Fedoroff NV (1989) Maize transposable elements. In: Berg DE, Howe MM (eds) Mobile DNA. American Soc Microbiol, Washington DC, pp 375–411

Fuchs RL, Heeren RA, Gustafson ME, Rogan GJ, Bartnicki DE, Leimgruber RM, Finn RF, Hershman A, Berberich SA (1993) Safety assessment of the neomycin phosphotransferase II (*npt*II) protein. Biotechnology 11:1537–1547

Galliano H, Muller AE, Lucht JM, Meyer P (1995) The transformation booster sequence is a retrotransposon derivative that binds to the nuclear scaffold. Mol Gen Genet 247:614–622

Gatz C, Lenk I (1998) Promoters that respond to chemical inducers. Trends Plant Sci 3:352–358

Gleave AP, Mitra DS, Mudge SR, Morris BAM (1999) Selectable marker-free transgenic plants without sexual crossing: transient expression of *cre* recombinase and use of a conditional lethal dominant gene. Plant Mol Biol 40:223–235

Goldsbrough AP, Lastrella CN, Yoder JI (1993) Transposition mediated re-positioning and subsequent elimination of marker genes from transgenic tomato. Biotechnology 11:1286–1292

Hoess RH, Wierzbicki A, Abemski K (1986) The role of the *lox*P-spacer region in P1 site-specific recombination. Nucleic Acid Res 5:2287–2300

Hooykaas PJJ, Schilperoort RA (1992) *Agrobacterium* and plant genetic engineering. Plant Mol Biol 18:15–38

Janssen B-J, Gardner RC (1989) Localised transient expression of GUS in leaf discs cocultivated with *Agrobacterium*. Plant Mol Biol 14:61–72

Jones JDG, Gilbert DE, Grady KL, Jorgensen RA (1987) T-DNA structure and gene expression in petunia plants transformed by *Agrobacterium tumefaciens* C58 derivatives. Mol Gen Genet 207:478–485

Jones JDG, Carland F, Lim E, Ralston E, Dooner HK (1990) Preferential transposition of the maize element *Activator* to linked chromosomal locations in tobacco. Plant Cell 2:701–707

Jorgenson R, Snyder C, Jones JDG (1987) T-DNA is organised predominantly in inverted repeat structures in plant transformed with *Agrobacterium tumefaciens* C58 derivatives. Mol Gen Genet 207:471–477

Kilby NJ, Snaith MR, Murray JAH (1993) Site-specific recombinases: tools for genome engineering. Trends Genet 9:413–421

Kilby NJ, Davies GJ, Snaith MR, Murray JAH (1995) FLP recombinase in transgenic plants: constitutive activity in stably transformed tobacco and generation of marked cell clones in *Arabidopsis*. Plant J 8:637–652

Kohli A, Leech M, Vain P, Laurie DA, Chrisyou P (1998) Transgene organisation in rice engineered through direct DNA transfer supports a two-phase integration mechanism mediated by the establishment of integration hot spots. Proc Natl Acad Sci USA 95:7203–7208

Komari T, Hiei Y, Saito Y, Murai N, Kumashiro T (1996) Vectors carrying two separate T-DNAs for cotransformation of higher plants mediated by *Agrobacterium tumefaciens* and segregation of transformants free from selection markers. Plant J 10:165–174

Lloyd AN, Davis RW (1994) Functional expression of the yeast FLP/FRT site-specific recombination system in *Nicotiana tabacum*. Mol Gen Genet 242:653–657

Lyznik LA, Mitchell JC, Hirayama L, Hodges TK (1993) Activity of yeast FLP recombinase in maize and rice protoplasts. Nucleic Acid Res 21:969–975

Lyznik LA, Hirayama L, Rao KV, Abad A, Hodges TK (1995) Heat-inducible expression of *FLP* gene in maize cells. Plant J 8:177–186

Lyznik LA, Rao KV, Hodges TK (1996) FLP-mediated recombination of FRT sites in the maize genome. Nucleic Acid Res 24:3784–3789

Ma JK, Hiatt A, Hein M, Vine ND, Wang F, Stabila P, van Dolleweerd C, Mostov K, Lehner T (1995) Generation and assembly of secretory antibodies in plants. Science 268:716–719

Maeser S, Kahmann R (1991) The Gin recombinase of phage Mu can catalyse site-specific recombination in plant protoplasts. Mol Gen Genet 230:170–176

McKnight TD, Lillis MT, Simpson RB (1987) Segregation of genes transferred to one plant cell from two separate *Agrobacterium* strains. Plant Mol Biol 8:439–445

Meyer P, Saedler H (1996) Homology-dependent gene silencing in plants. Ann Rev Plant Physiol Plant Mol Biol 47:23–48

Odell J, Russell SH (1994) Use of site-specific recombination systems in plants. In: Paszkowski J (ed) Homologous recombination and gene silencing in plants. Kluwer, Dordrecht, pp 219–270

Odell J, Caimi P, Sauer B, Russell S (1990) Site-directed recombination in the genome of transgenic tobacco. Mol Gen Genet 223:369–378

Onouchi H, Yokoi K, Machida C, Matsuzaki H, Oshima Y, Matsuoka K, Nakamura K, Machida Y (1991) Operation of an efficient site-specific recombination system of *Zygosaccharomyces rouxii* in tobacco cells. Nucleic Acid Res 19:6373–6378

Onouchi H, Nishihama R, Kudo M, Machida Y, Machida C (1995) Visualisation of site-specific recombination catalysed by a recombinase from *Zygosaccharomyces rouxii* in *Arabidopsis thaliana*. Mol Gen Genet 247:653–660

Ow DW, Medberry SL (1995) Genome manipulation through site-specific recombination. Crit Rev Plant Sci 14:239–261

Puchta H, Swoboda P, Gal S, Blot M, Hohn B (1995) Somatic intrachromosomal homologous recombination events in populations of plant siblings. Plant Mol Biol 28:281–292

Russell SH, Hoopes JL, Odell JT (1992) Directed excision of a transgene from the plant genome. Mol Gen Genet 234:49–59

Slater S, Mitsky TA, Houmiel KL, Hao M, Reiser SE, Taylor NB, Tran M, Valentin HE, Rodriguez DJ, Stone DA, Padgette SR, Kishore G, Gruys KJ (1999) Metabolic engineering of *Arabidopsis* and *Brassica* for poly (3-hydroxybutyrate-*co*-3-hydroxyvalerate) copolymer production. Nat Biotechnol 17:1011–1016

Smigocki AC, Owens LD (1989) Cytokinin-to-auxin ratios and morphology of shoots and tissues transformed by a chimeric isopentenyl transferase gene. Plant Physiol 91:808–811

Spencer TM, O'Brien JV, Start WG, Adams TR, Goron-Kamm WJ, Lemaux PG (1992) Segregation of transgenes in maize. Plant Mol Biol 18:210–210

Spielmann A, Simpson RB (1986) T-DNA structure in transgenic tobacco plants with multiple independent integration sites. Mol Gen Genet 205:34–41

Srivastava V, Anderson OD, Ow DW (1999) Single-copy transgenic wheat generated through the resolution of complex integration patterns. Proc Natl Acad Sci USA 96:11117–11121

Stougaard J (1993) Substrate-dependent negative selection in plants using a bacterial cytosine deaminase gene. Plant J 3:755–761

Sugita K, Matsunaga E, Ebinuma H (1999) Effective selection system for generating marker-free transgenic plants independent of sexual crossing. Plant Cell Rep 18:941–947

Sugita K, Kasahara T, Matsunaga E, Ebinuma H (2000) A transformation vector for the production of marker-free transgenic plants containing a single copy transgene at high frequency. Plant J 22:461–469

Surin BP, De Feyter RC, Graham MW, Waterhouse PM, Keese PK, Shahjahan A (1997) Single-step excision means. Patent WO 97/37012

Wakita Y, Otani M, Iba K, Shimada T (1998) Co-integration, co-expression and co-segregation of an unlinked selectable marker gene and NtFAD3 gene in transgenic rice plants produced by particle bombardment. Genes Genet Syst 73:219–226

Yoder JI, Goldsbrough AP (1994) Transformation systems for generating marker-free transgenic plants. Biotechnology 12:263–267

Zechendorf B (1994) What the public thinks about biotechnology. Biotechnology 12:870–875

Zubko E, Scutt C, Meyer P (2000) Intrachromosomal recombination between *att*P regions as a tool to remove selectable marker genes from tobacco transgenes. Nat Biotechnol 18:442–445

7 GST-MAT Vector for the Efficient and Practical Removal of Marker Genes from Transgenic Plants

H. Ebinuma, K. Sugita, E. Matsunaga, S. Endo, and K. Yamada

7.1 Introduction

In current transformation systems, a selectable marker gene is co-delivered with the gene of interest to identify and separate rare transgenic cells from non-transgenic cells. Since, during transformation, only a few plant cells accept the integration of foreign DNA, most of the cells remain non-transgenic. Usually, conditional dominant genes, which have no influence on the growth or morphology of plants, are used as selectable markers because they remain in the transgenic plants after transformation. The corresponding selective agents, which inhibit the growth of non-transgenic cells, are applied to the culture medium to identify transgenic plants. However, these selection systems have three potential pitfalls. (1) The negative effects of selective agents decrease the ability of transgenic cells to proliferate and differentiate into transgenic plants. (2) The presence of marker genes in transgenic plants precludes the use of the same marker gene for gene stacking through re-transformation. (3) The recent public concerns regarding the release of antibiotic-resistance genes limit their use for the commercialization of transgenic crops.

The removal of selectable marker genes is a reasonable strategy for addressing gene stacking and public concerns. However, the feasibility of practical removal systems depends on the economic value of marker-free transgenic plants, and the costs associated with their production. Several approaches to remove a selectable marker gene from transgenic plants have been reported (review: Ebinuma et al. 2001). These transformation systems consist of a site-specific recombination system (Cre/*lox*) or a phage-attachment region (*attP*) to remove the selectable marker gene, and a transposable element system (*Ac*) or a co-transformation system to segregate the gene of interest from the selectable marker gene. In these systems, antibiotic or herbicide resistance genes are used to select transgenic plants. Once these selectable marker genes are eliminated from transgenic cells by the removal systems, selective agents can not be used to select non-chimeric marker-free transgenic plants. Therefore, a two-step transformation procedure is needed to generate marker-free transgenic plants. First, transgenic plants containing selectable marker genes are selected by using an antibiotic or herbicide, and the selectable marker genes are then removed or separated from the transgenic plants. DNA analysis is used to identify marker-free transgenic plants. Overall, the process is more time-consuming than conventional transformation methods.

Molecular Methods of Plant Analysis, Vol. 22
Testing for Genetic Manipulation in Plants
Edited by J.F. Jackson, H.F. Linskens, and R.B. Inman
© Springer-Verlag Berlin Heidelberg 2002

Recently, removal systems combined with a positive marker, which are called MAT vectors, have been developed to address the three pitfalls mentioned above (negative effects, gene stacking, public concerns) (review: Ebinuma et al. 1997b, 2000; Ebinuma and Komamine 2001). In this chapter, we summarize a transformation procedure using *ipt*-type MAT vectors and introduce their application to tobacco, hybrid aspen and rice.

7.2 *ipt*-Type MAT Vectors

In *ipt*-type MAT vectors, the isopentenyl transferase (*ipt*) gene of *A. tumefaciens* PO22 is used as a selectable marker to regenerate transgenic plants and to select non-chimeric marker-free transgenic plants (Wabiko et al. 1989). The *ipt* gene codes the isopentenyl transferase that catalyzes cytokinin synthesis (Akiyoshi et al. 1984; Barry et al. 1984) and causes the proliferation of transgenic cells and the differentiation of adventitious shoots. The *ipt* gene is not commonly used for transformation because transgenic plants that contain the *ipt* gene exhibit seriously abnormal phenotypes and do not show apical dominance or rooting ability. Therefore, the *ipt*-type MAT vectors are designed to remove the *ipt* gene from the transgenic plants after transformation by using the *Ac* transposable element (pNPI106) or the site-specific recombination *R/RS* system (pNI132), and to recover the normal phenotype (Fig. 7.1).

7.2.1 Transposable Element

In the pNPI106 vector, the chimeric *ipt* gene with a 35S promoter is inserted into the *Ac* transposable element to remove it from transgenic cells after transformation. The modified *Ac* cassette is used for selection, and is called the "hit and run" cassette of the MAT vector. The MAT cassette is inserted into the *Sse*I site of the binary vector plasmid pBI121 (CLONTECH). Both the *npt*II and *gus*A genes outside of the MAT cassette are used as model genes of interest (Fig. 7.2A). *Ac* is a maize transposable element that can transpose to new locations within a genome (Fedoroff 1989). In the transposition process, about 10% of the excised *Ac* elements do not reinsert and therefore disappear, or reinsert into a sister chromatid that is subsequently lost by somatic segregation (Belzile et al. 1989). As the 35S-*ipt* gene is inserted into the *Ac* element in the MAT vector, it is lost along with the *Ac* element. We infected leaf segments of tobacco (*Nicotiana tabaccum* cv. SR1) plants and stem segments of hybrid aspen (*Populus sieboldii* × *P. grandidentata*) plants with *Agrobacterium* containing the binary vector pNPI106 (Ebinuma et al. 1997a). Of 100 adventitious tobacco shoots regenerated in kanamycin- and hormone-free MS medium, 63 (63%) showed the *ipt*-shooty phenotype that had lost apical dominance and rooting ability due to the overproduction of cytokinin. Within 6 months after infection,

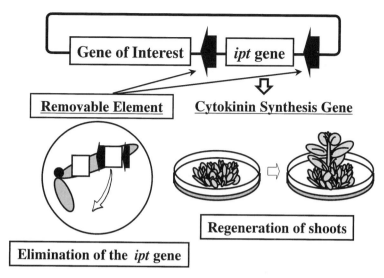

Fig. 7.1. Principle of the *ipt*-type MAT vectors. The *ipt*-type MAT vectors combine the removal system with the *ipt* gene that catalyzes cytokinin synthesis. The cassette is used for both the regeneration of transgenic plants and selection of marker-free transgenic plants

several normal shoots that exhibited normal apical dominance appeared from three of the 63 (4.8%) *ipt*-shooty lines (Fig. 7.1). When used for the transformation of hybrid aspen, six normal shoots appeared from three of 20 (15%) *ipt*-shooty lines. Molecular analyses showed that these normal tobacco and aspen plants contained the *npt*II and *gus*A genes, but had lost the 35S-*ipt* gene. These results indicate that the *Ac* element can remove a selectable marker gene from transgenic tobacco and hybrid aspen plants without sexual crossing, and the chimeric *ipt* gene can be efficiently used as a visible marker to select marker-free transgenic tobacco and hybrid aspen plants.

7.2.2 Site-Specific Recombination System

In the pNPI132 vector, the chimeric *ipt* gene with a 35S promoter is combined with the site-specific recombination *R*/RS system to remove them from the transgenic cells after transformation. The 35S-*ipt* gene and the recombinase (*R*) gene with a 35S promoter are placed within two directly oriented recognition sites (RS). The *R*/RS system cassette is used for selection, and is called the "hit and run" cassette of the MAT vector. The MAT cassette is inserted into the *Sse*I site of the binary vector plasmid pBI121. Both the *npt*II and *gus*A genes outside the MAT cassette are used as model genes of interest (Fig. 7.2A). The *R*/RS system was isolated from the circular plasmid pSR1 of *Zygosaccharomyces rouxii*. In this recombination system, the DNA fragment between the

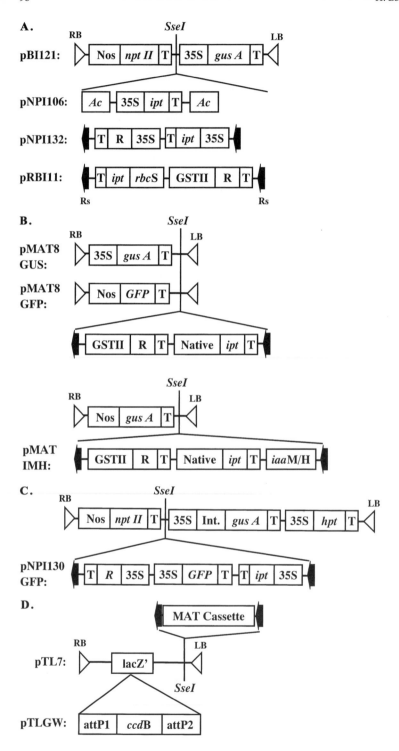

two directly oriented RS is excised from the plant genome using the *R* gene product (recombinase) (Onouchi et al. 1991) (Fig. 7.3). We describe the transformation of tobacco using pNPI132 as follows (Sugita et al. 1999) (Fig. 7.4):

1. Gene transfer by infection. Eighty-two pieces of leaf segments of tobacco (*Nicotiana tabaccum* cv. SR1) plants were infected with *A. tumefaciens* containing the pNPI132 vector and co-cultivated on hormone-free MS agar medium containing 50 mg/l acetosyringone for 3 days.

2. Regeneration of transgenic plants. The explants were transferred to hormone-free MS agar medium containing 500 mg carbenicillin/l but not kanamycin (nonselective medium). One month after infection, 134 regenerated adventitious buds were separated from the leaf segments and transferred to the same medium. After 1 month of cultivation, we visually classified these developed buds into three distinct phenotypes: 68 normal shoots, 18 moderate *ipt*-shooty, and 48 extreme *ipt*-shooty phenotypes. These abnormal shoots (49.3%) had lost apical dominance and rooting ability due to the overproduction of cytokinin.

3. Appearance of marker-free plants. After 1 month of further cultivation, normal shoots exhibiting apical dominance appeared from 17 of 18 moderate *ipt*-shooty clones. All of the 48 extreme *ipt*-shooty clones were subcultured on the same fresh medium to monitor the appearance of normal shoots. Normal shoots appeared from 10 of 48 extreme *ipt*-shooty clones within 4 months after infection. After 4 months more of cultivation, 22 additional extreme *ipt*-shooty clones developed multiple normal shoots. These shoots were transferred to the same medium, grew normally and rooted.

4. DNA analysis of the normal plants. Of 68 normal shoots, 16 developed directly from the adventitious shoots, 17 normal shoots appeared from moderate *ipt*-shooty clones and 32 normal shoots appeared from extreme *ipt*-shooty clones, and all were subjected to PCR analysis. The predicted 0.8-

---

Fig. 7.2A–D. MAT vectors. **A** pNPI106, pNPI132 and pRBI11 have a MAT cassette composed of the 35S-*ipt* gene and *Ac* element, the 35S-*ipt* and 35S-*R* genes, and the *rbc*S 3B-*ipt* and GST-II-27-*R* genes, respectively. These cassettes are inserted into a *Sse*I site of pBI121. **B** pMAT8 has a MAT cassette composed of the native *ipt* and GST-II-27-*R* genes. The cassette is inserted into a *Sse*I site of pTL7 containing the 35S-*gus*A and Nos-*GFP* genes to create pMAT8GUS and pMAT8GFP, respectively. pMAT IMH has a MAT cassette composed of the native *ipt*, GST-II-27-*R* and *iaa* M/H genes. The cassette is inserted into a *SSe* I site of pTL7 containing the Nos-*gus* A gene. **C** pNPI130GFP has a MAT cassette composed of the 35S-*ipt*, 35S-*GFP* and 35S-*R* genes. The cassette is inserted into a *Sse*I site of pBI121 containing 35S-*hpt* gene. The *gus*A has an intron sequence. **D** pTL7 has lacZ′ multi-cloning sites and a *Sse*I site for insertion of a gene of interest and a MAT cassette, respectively. pTLGW has a *ccd*B gene flanked by the attP1 and attP2. The MAT cassettes are flanked by two directly oriented RS sequences. The *npt*II, *hpt* and *gus*A genes outside of the cassette are used as models for a gene of interest. *ipt*: Isopentenyl transferase gene *iaa* M/H: tryptophan monooxgenase and indoleacetamide hydrolase genes, *R* recombinase gene, *RS* recognition sequence, *npt*II neomycin phosphotransferase gene, *hpt* hygromycin phosphotransferase gene, *gus*A β-glucuronidase gene, *35S-P*: CaMV³⁵S promoter, *rbcS-P rbc*S 3B promoter, *GSTII-P* GST-II-27 promoter, *T* nopaline synthase terminator, *RB* and *LB*: right and left border sequences of a T-DNA, respectively

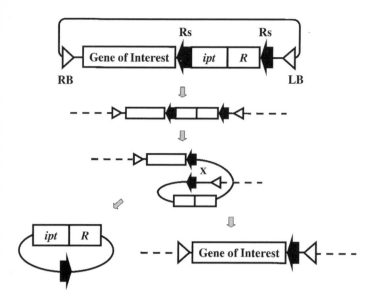

Fig. 7.3. Site-specific recombination system (*R*/Rs). A gene of interest is joined to the RS-flanked *ipt* and *R* genes. Recombinase (*R*) catalyzes recombination between two directly oriented recognition sites (Rs) and produce a circular DNA containing a MAT cassette. The cassette eliminates from a plant genome, and a gene of interest and one Rs sequence remains

(1) Two-step transformation

(2) Single-step transformation

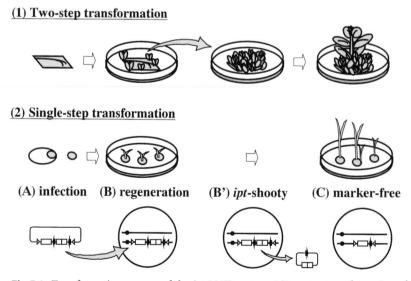

(A) infection   (B) regeneration   (B') *ipt*-shooty   (C) marker-free

Fig. 7.4. Transformation process of the *ipt*-MAT vectors. *1* Two-step transformation of tobacco plants using leaf-discs. *2* Single-step transformation of rice plants using scutellum tissues. *A* Gene transfer by infection of *Agrobacterium*, *B* regeneration of transgenic shoots, *B'* formation of the *ipt*-shooty phenotypes, *C* appearance of marker-free transgenic plants

kb *ipt* fragment and 1.7-kb *gus*A fragment were amplified with the primer pairs IPT1-IPT2 and GUS1-GUS2, respectively. The predicted 3.2-kb fragment could be amplified with the primer pair EX1-EX2 if the "hit and run" cassette was excised. In 2 (13%) of 16 normal shoots from normal clones, 7 (39%) of 17 normal shoots from moderate *ipt*-shooty clones and 32 (100%) of 32 normal shoots from extreme *ipt*-shooty clones, a predicted *ipt* fragment was not amplified but an excision fragment was amplified by PCR analysis. All of the 41 normal shoots were kanamycin-resistant in a kanamycin assay, while six of 7 normal shoots from moderate *ipt*-shooty clones and 20 of 32 normal shoots from extreme *ipt*-shooty clones had GUS activity in a GUS assay. These results indicate that 41 normal shoots were transgenic plants that did not contain the marker genes.

Using the pNPI106 vector, we obtained marker-free transgenic tobacco plants from only three (5%) of 63 *ipt*-shooty clones 8 months after infection with *Agrobacterium* (Ebinuma et al. 1997a). However, we regenerated marker-free transgenic tobacco plants from seven (39%) of 18 moderate *ipt*-shooty clones and 32 (70%) of 48 extreme *ipt*-shooty clones within 8 months after infection using pNPI132 (Sugita et al. 1999). These results indicate that the *ipt*-type MAT vector enables the generation of marker-free transgenic plants without sexual crossings and that the *R/RS* system is more practical for removal of a selectable marker gene than the *Ac* element.

7.2.3 Advantages of the *ipt* Gene

The generation of marker-free transgenic plants using *ipt*-type MAT vectors consists of three characteristic processes.

1. Regeneration of transgenic plants. In current transformation systems, plant growth regulators are exogenously applied to the culture medium to stimulate the regeneration of transgenic plant cells. However, since they also promote the regeneration of non-transgenic cells, selective agents are needed to inhibit their regeneration. When *ipt*-type MAT vectors are used for transformation, the *ipt* gene overproduces cytokinin in transgenic cells and differentiates them preferentially instead of plant growth regulators. Transgenic plants exhibit the *ipt*-shooty phenotype which has lost apical dominance and rooting ability. Therefore, we can identify transgenic plants without using selective agents.

2. Removal of a selectable marker gene. In current systems for the generation of marker-free transgenic plants, antibiotic or herbicide resistance genes are used as a selectable marker gene and combined with removal systems (review: Ebinuma et al. 2001). When the excision events of a removal system occur during transformation, a selectable marker gene is eliminated from the genome of transgenic cells. Usually, since transgenic cells contain several copies of inserts in the genome, three kinds of transgenic cells appear

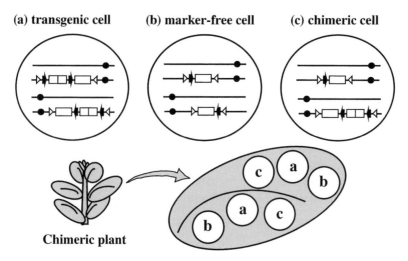

(a) transgenic cell   **(b) marker-free cell**   **(c) chimeric cell**

Chimeric plant

Fig. 7.5a–c. Generation of chimeric transgenic plants. Three kinds of transgenic cells appear during transformation. **a** All inserts have a marker gene, **b** all inserts lose it, **c** several inserts have it and the others lose it

during the removal of selectable marker genes (Fig. 7.5): (a) Transgenic cells in which a selectable marker gene remains in all copies of inserts. (b) Marker-free transgenic cells in which a selectable marker gene is eliminated from all copies of inserts. (c) Chimeric transgenic cells in which a selectable marker gene remains in several copies of inserts but is eliminated from the other copies. Therefore, chimeric transgenic plants in which three kinds of transgenic cells co-exist appear during transformation.

3. Separation of marker-free transgenic plants. Usually, once a selectable marker gene is removed from transgenic cells, selective agents can not be used to identify marker-free transgenic cells, since they are killed in culture medium containing selective agents. Therefore, chimeric transgenic plants are crossed to segregate non-chimeric marker-free transgenic plants at their progeny, which are identified by DNA analysis. In the *ipt*-MAT vectors, once the *ipt* gene is removed from transgenic plants exhibiting *ipt*-shooty phenotypes, marker-free transgenic shoots can recover apical dominance and rooting ability, and extend from the *ipt*-shooty clones. Chimeric plants also appear, but exhibit the *ipt*-shooty phenotype because the remaining *ipt* gene in several cells overproduces cytokinin. Therefore, transgenic plants that exhibit a normal phenotype are identified as marker-free transgenic plants that only contain cells in which the *ipt* gene is eliminated from all copies of inserts. Therefore, neither DNA analysis nor sexual crossing is needed to identify non-chimeric marker-free transgenic plants.

7.3 Two-Step Transformation

7.3.1 Promoter of the *R* Gene

In two-step transformation methods using the *ipt*-type MAT vector, first, transgenic plants are selected as the *ipt*-shooty phenotype, and marker-free transgenic plants that appear from them are then separated (Fig. 7.4). The *ipt*-type MAT vector pNPI132 uses a 35S promoter to control the expression of the *R* gene. However, we observed the excision events of MAT cassettes before shoot formation because the 35S promoter is expressed constitutively. To control excision events, we changed a 35S promoter of the *R* gene to a chemically inducible promoter of the glutathione-S-transferase (GST-II-27) gene from *Zea mays*. The herbicide antidote safener has been reported to induce the expression of the GST-II-27 promoter (Holt et al. 1995). The GST-MAT vector pMAT8 has the *ipt* gene with the native promoter and the GST-*R* gene (Fig. 7.2B). The GUS gene with the 35S promoter is inserted into the lacZ′ multicloning sites of pMAT8 as a model gene of interest.

The production of marker-free transgenic plants using pMAT8:35SGUS involves four main steps (Sugita et al. 2000a: Fig. 7.6).

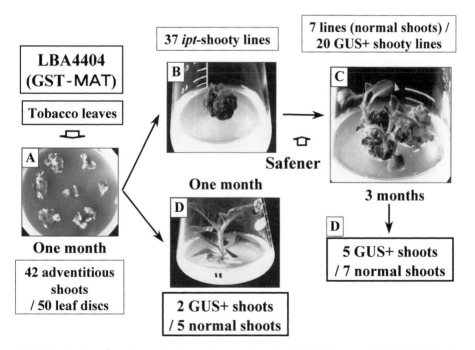

Fig. 7.6A–D. Transformation of tobacco plants using the GST-MAT vector (pMAT8:GUS). **A** Regeneration of adventitious shoots from leaf segments; **B** an *ipt*-shooty phenotype lacking apical dominance and rooting ability; **C** appearance of normal shoots exhibiting apical dominance from *ipt*-shooty clones; **D** a marker-free transgenic plant (MF)

1. Gene transfer by infection. Sixty pieces of leaf segments of tobacco (*Nicotiana tabaccum* cv. SR1) plants were infected with *A. tumefaciens* containing the pMAT8:35SGUS vector and co-cultivated on hormone-free MS agar medium containing 50 mg acetosyringone/l for 3 days.
2. Regeneration of transgenic plants. The explants were transferred to hormone-free MS agar medium containing 500 mg carbenicillin/l, but not kanamycin (nonselective medium). One month after infection, 42 regenerated adventitious buds were separated from the leaf segments and transferred to the same medium. After 1 month of cultivation, we visually classified these developed buds into two distinct phenotypes: 5 normal shoots and 37 *ipt*-shooty phenotypes. These abnormal shoots had lost apical dominance and rooting ability due to the overproduction of cytokinin.
3. Appearance of marker-free plants. Twenty GUS-positive *ipt*-shooty clones were subcultured monthly to the same fresh medium and safener-induction MS medium containing 30 mg/l safener (R29148). Several normal shoots appeared from 7 of 20 *ipt*-shooty clones within 3 months of induction with safener. These shoots were transferred to the same medium, grew normally and rooted.
4. DNA analysis of normal plants. All 12 normal shoots were subjected to PCR analysis. A predicted *ipt* fragment was not amplified in two of five (40%) normal shoots that developed directly from the adventitious shoots, and five of seven7 (70%) normal shoots appeared from seven *ipt*-shooty clones, but an excision fragment was amplified by PCR analysis. Seven of 12 (58%) normal shoots were marker-free transgenic plants and five (42%) normal shoots were non-transgenic escapes. These results show that we can both rapidly produce marker-free transgenic plants without the production of *ipt*-shooty intermediates and induce the generation of marker-free transgenic plants using safener. We investigated the copy number of the GUS gene in these seven marker-free transgenic plants by Southern blot analysis, and found that six of seven (86%) marker-free transgenic tobacco plants had a single GUS gene, while one plant had two genes. These results indicate that the GST-MAT vector is useful for producing marker-free transgenic plants containing a single-copy transgene at high frequency.

Using pNPI106, we obtained three marker-free transgenic tobacco plants that contained one or two inserted copies of T-DNA (Ebinuma et al. 1997a). If more than one expressed copy of the *ipt* gene were inserted into the plant genome of the transgenic shoots, the elimination of one copy would not cause a loss of *ipt* function (*ipt*-shooty). This inference leads to the expectation that marker-free transgenic plants will be derived from low-copy-number transgenic plants. In contrast, 67% of marker-free transgenic tobacco plants had more than three inserted copies of T-DNA (one copy: 10 plants, 2 copies: 2 plants, over 3 copies: 24 plants) using pNPI132 (Sugita et al. 1999). These results indicate that the R/RS system driven by the 35S promoter is more active than

Ac. The 35S-*R* gene might be expressed in the transgenic callus and remove the *ipt* gene before the transgenic shoots are regenerated. Therefore, the transgenic callus containing a low copy number of the inserted *ipt* gene might lose *ipt* activity for regenerating shoots. We also observed that the GST-II-27 promoter was slightly active in the absence of safener and its activity was greatly enhanced after safener induction (data not shown). The GST-*R* gene might not be strongly expressed in the callus, and such expression may be induced after shoot formation by safener. Therefore, we could successfully increase both the regeneration efficiency of *ipt*-shooty transgenic plants and the generation efficiency of marker-free transgenic plants containing one copy of the inserts by using pMAT8. Transgenic plants with a low copy number are more valuable than those with a high copy number for practical use. Therefore, the GST-II-27 promoter is preferable for controlling excision events during transformation using the *ipt* gene.

7.3.2 Promoter of the *ipt* Gene

During transformation using the *ipt*-type MAT vector, non-transgenic shoots are regenerated together with transgenic shoots, since the overproduction of cytokinin by the *ipt* gene causes it to leak out from transgenic cells and promote the regeneration of non-transgenic cells around the transgenic cells. Selective agents and the corresponding resistance genes are commonly used to avoid the regeneration of non-transgenic plants during transformation. In the case of the *ipt*-type MAT vector, by using the *ipt* gene, transgenic plants are visually identified as the *ipt*-shooty phenotype and selected. However, further cultivation of regenerated buds is needed to avoid non-transgenic plants that escape visual selection because complete visual selection of transgenic plants is difficult to achieve just after the regeneration of adventitious buds. To increase the percentage of transgenic shoots relative to non-transgenic shoots, we fused the *ipt* gene with several different promoters to optimize the cytokinin level required for the regeneration of transgenic cells.

pIPT5, 10 and 20 are derivatives of the binary vector pBI121 which contain the chimeric *ipt* gene with a 35S promoter, a native *ipt* promoter and a *rbc*S 3B promoter, respectively. The native *ipt* promoter can be isolated from the Ti plasmids of *A. tumefaciens* PO22 (Wabiko et al. 1989). We observed that this promoter was active in the shoots and roots and that its activity was increased by wounding (data not shown). The *rbc*S 3B promoter was isolated from tomato by PCR. It is induced by light and active only in green tissues (Sugita and Gruissem 1987).

Leaf segments of tobacco (*Nicotiana tabaccum* cv. SR1) and stem segments of hybrid aspen (*Populus sieboldii* x *P. grandidentata*) were infected with *A. tumefaciens* containing pIPT5, 10, and 20 or pBI121, and co-cultivated on hormone-free MS agar medium containing 40 mg acetosyringone/l for 2 days and on hormone-free modified MS agar medium (800 mg ammonium nitrate/l,

2 g potassium nitrate/l) containing 40 mg acetosyringone/l for 3 days, respectively. When we used pIPT5, 10 and 20, the explants of tobacco and hybrid aspen were transferred to the same hormone-free medium containing 500 mg carbenicillin/l. The regenerated adventitious buds of tobacco were separated from the leaf segments and transferred to fresh MS medium containing 500 mg carbenicillin/l after 1.5 months of cultivation, and the adventitious buds of hybrid aspen from stem segments were transferred to fresh modified MS medium containing 500 mg carbenicillin/l after 2.5 months of cultivation. After 1 month more of cultivation, we visually classified the developed buds into the normal and *ipt*-shooty phenotypes, and subjected them to GUS assay. When we used pBI121, the explants of tobacco and hybrid aspen were transferred to MS (0.1 mg naphthylacetic acid/l, 1 mg benzylaminopurine/l) and modified MS (0.5 mg zeatin/l) medium containing 500 mg carbenicillin/l and 100 mg kanamycin/l, respectively. After 2 months of cultivation, regenerated adventitious buds of tobacco were separated from the leaf segments and transferred to hormone-free MS medium containing 500 mg carbenicillin/l and 100 mg kanamycin/l, and after 4 months of cultivation, the adventitious buds of hybrid aspen from stem segments were transferred to hormone-free modified MS medium containing 500 mg carbenicillin/l and 100 mg kanamycin/l. After 1 month of cultivation, we subjected these developed shoots to kanamycin and GUS assay, and classified them into kanamycin-resistant transgenic shoots and non-transgenic shoots.

When we used pIPT5, 10, and 20 and pBI121 for the transformation of tobacco, we obtained two to three times more transgenic shoots (*ipt*-shooty) per leaf segment using pIPT5, 10 and 20 than with pBI121. Interestingly, pIPT20 showed higher ratios of transgenic shoots to regenerated shoots and of transgenic shoots with GUS activity to transgenic shoots than pIPT5 and 10. When we used pIPT5, 10 and 20 for the transformation of hybrid aspen, we obtained four to six times more transgenic shoots (*ipt*-shooty) per stem segment with pIPT20 than with pIPT5 and 10. Interestingly, regenerated shoots were mostly non-transgenic using pIPT5 (11.1%) and 10 (13.7%), in contrast to pIPT20 (53.1%). In addition, the ratio of transgenic shoots with GUS activity to transgenic shoots using pIPT20 (86.8%) was greater than that using pIPT5 (62.1%) and 10 (75.0%) (Table 7.1). The transformation efficiency using pIPT20 was tenfold greater than that with pBI121 (3.86/0.35) and the culture time was shortened by nearly half (2.5/4).

These results show that the chimeric *ipt* gene with a *rbc*S 3B promoter increases the percentage of transgenic shoots relative to non-transgenic shoots more in hybrid aspen than in tobacco. In hybrid aspen, adventitious buds were regenerated only from the green parts of the callus. The *rbc*S 3B-*ipt* gene would actively express and produce cytokinin only in a green callus but not in a white callus. Therefore, we believe that the cytokinin level in transgenic green calli might be high enough to differentiate transgenic shoots, while that of transgenic white calli may be too low to regenerate the non-transgenic cells around them.

Table 7.1. Transformation efficiency of hybrid aspens using the chimeric *ipt* genes. pIPT5 (35S-*ipt*), pIPT10 (native *ipt*), pIPT20 (*rbcS-ipt*): two and a half months after infection, regenerated adventitious shoots were separated from stem segments and placed on hormone-free modified MS agar medium containing 500 mg carbenicillin/l, pBI121: 4 months after infection on modified MS agar medium containing 0.5 mg zeatin/l, 500 mg carbenicillin/l and 100 mg kanamycin/l

Constructs	Number of explants	Number of transgenic shoots per stem segment	Percentage of transgenic shoots in regenerated shoots	Percentage of transgenic shoots with GUS activity
pIPT5	49	0.67	11.1	62.1
pIPT10	40	1.00	13.7	75.0
pIPT20	64	3.86	53.1	86.8
pBI121	60	0.35	100	85.7

We constructed the GST-MAT vector pRBI11 composed of the *ipt* gene with the *rbc*S 3B promoter and the *R* recombinase gene with the GST-II-27 promoter (Fig. 7.2A). The transformation process of hybrid aspen using pRBI11 is as follows (Fig. 7.7):

1. Gene transfer by infection. Thirty and 20 stem segments of hybrid aspen (*Populus sieboldii* × *P. grandidentata*) were infected with *A. tumefaciens* containing the pRBI11 vector and co-cultivated on hormone-free modified MS agar medium (800 mg ammonium nitrate/l, 2 g potassium nitrate/l) containing 40 mg acetosyringone/l for 3 days.
2. Regeneration of transgenic plants. The explants were transferred to the same hormone-free medium containing 500 mg carbenicillin/l without kanamycin (nonselective medium). After two and a half months of cultivation, the stem segments, together with regenerated adventitious buds, were transferred to the same medium. After 1 month of cultivation, about half of the regenerated shoots exhibited the *ipt*-shooty phenotype. Without their separation and further cultivation, the transgenic shoots with the *ipt* gene could be clearly identified as the *ipt*-shooty phenotype. GUS activity was detected in about 80% of the *ipt*-shooty clones regenerated from 23 (46.0%) of 50 stems by GUS assay.
3. Appearance of marker-free plants. We previously reported that marker-free transgenic tobacco could be obtained by safener induction (Sugita et al. 2000a). However, we observed that, compared to tobacco plants, it was much more difficult to generate marker-free transgenic hybrid aspen plants because of the low regeneration frequency and high degree of damage with safener. Furthermore, we observed that the GUS gene with the GST-II-27 promoter was induced by wounding and actively expressed at cut sites of the stem segments of hybrid aspen (data not shown). Therefore, instead of safener induction, we independently cut 14 GUS-positive *ipt*-shooty clones into small pieces and transferred them to a modified MS medium containing 0.5 mg zeatin/l and 500 mg carbenicillin/l (shoot-inducing medium).

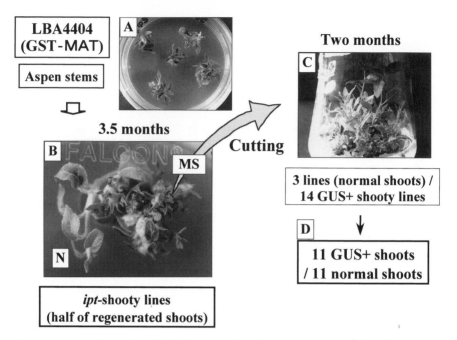

Fig. 7.7A–D. Transformation of hybrid aspens using the GST-MAT vector (pRBI11). A Regeneration of adventitious shoots from stem segments; **B** normal phenotypes (*N*) and *ipt*-shooty phenotypes (*MS*) lacking apical dominance and rooting ability; **C** appearance of normal shoots exhibiting apical dominance from *ipt*-shooty clones; **D** a marker-free transgenic plant (*MF*)

Eleven normal shoots appeared from 3 of 14 (21.4%) *ipt*-shooty clones within 2 months of induction by wounding. These shoots were transferred to 2/3 MS medium containing 0.05 mg/l IBA (root-inducing medium), in which they grew normally and rooted. Kanamycin resistance and GUS activity of 11 normal shoots were confirmed by kanamycin and GUS assays.

4. DNA analysis of the normal plants. We subjected eight developed shoots from three *ipt*-shooty clones to PCR analysis. In all eight normal plants, a predicted *ipt* fragment was not amplified, but an excision fragment was amplified by PCR analysis. All eight normal shoots were marker-free transgenic plants and no non-transgenic escapes were observed. These results indicate that we can produce transgenic aspen plants at high frequency and induce the generation of marker-free transgenic aspen plants by wounding.

7.3.3 Combination of the *ipt* and *iaa*M/H Genes

In current transformation methods, two kinds of plant growth regulators (cytokinin, auxin) are applied to the tissue culture medium for the regeneration of transgenic plants. Although the endogenous hormone levels are very different among plant species as well as among plant tissues, it is well established that adventitious shoots regenerate from a plant tissue culture with an increase in the amount of cytokinin relative to that of auxin in the culture medium. In the *ipt*-type MAT vector, the *ipt* gene is used for the regeneration of transgenic plants instead of plant growth regulators (Endo et al. 2001). It has also been reported that the overproduction of cytokinin by *ipt* genes can cause the regeneration of transgenic shoots in potato (Ooms et al. 1983), cucumber (Smigocki and Owens 1989), tobacco (Medford et al. 1989; Schmulling et al. 1989; Smigocki and Owens 1989; Estruch et al. 1991; Smart et al. 1991; Li et al. 1992), *Arabidopsis* (Medford et al. 1989), strawberry (James et al. 1990), peach (Smigocki and Hammerschlag 1991), poplar (Schwartzenberg et al. 1994) and basket willow (Vahala et al. 1993). However, the control of both cytokinin and auxin is preferable to optimizing the hormone levels in plant tissue and regenerating transgenic shoots in many plant species. Therefore, we combined the *ipt* genes with the *iaa*M/H genes to manipulate both the auxin and cytokinin levels. The *iaa*M/H genes code for a tryptophan monooxgenase and an indoleacetamide hydrolase, which catalyze auxin synthesis (Thomashow et al. 1984). The *iaa*M/H genes are isolated from *A. tumefaciens* PO22 (Wabiko et al. 1989), which induces a large tumor on the trunks of hybrid aspen.

pIPTIMH are derivatives of the binary vector pBI121 which contain the native *ipt* and *iaa*M/H genes. We transformed tobacco (*Nicotiana tabaccum* cv. SR1) and poplar (*Populus tomentosa*) plants with *Agrobacterium* containing pIPT10 and pIPTIMH to compare the regenerative abilities of the *ipt* gene and the *ipt* gene combined with the *iaa*M/H genes. We observed that both the proliferation of calli and the differentiation of shoots were induced faster from tobacco leaf-discs and poplar stem segments infected with pIPTIMH than from those infected with pIPT10. Transgenic plants with pIPTIMH also exhibited *ipt*-shooty phenotypes which lacked apical dominance and rooting ability, and were easily distinguished from non-transgenic shoots. These results show that the *iaa*M/H genes promote both the proliferation of calli and the differentiation of shoots in combination with the native *ipt* gene, and the *ipt* gene combined with the *iaa*M/H gene can be used as a selectable marker to identify transgenic plants.

We constructed the GST-MAT vector pMATIMH composed of the native *ipt* and *iaa*M/H genes and the *R* gene with the GST-II-27 promoter (Fig. 7.2B). The production of marker-free transgenic plants using pIPTMH involves four main steps:

1. Gene transfer by infection. Twenty pieces of leaf segments of tobacco (*Nicotiana tabaccum* cv. SR1 plants) were infected with *A. tumefaciens* containing the MATIMH vector and co-cultivated on hormone-free MS agar medium containing 50 mg acetosyringone /l for 3 days.
2. Regeneration of transgenic plants. The explants were transferred to hormone-free MS agar medium containing 500 mg carbenicillin/l but not

kanamycin (nonselective medium). One month after infection, 20 regenerated adventitious buds were separated from the leaf segments and transferred to the same medium. After 1 month of cultivation, we visually identified 15 *ipt*-shooty phenotypes which lacked apical dominance and rooting ability.

3. Appearance of marker-free plants. Twelve GUS-positive *ipt*-shooty clones were transferred to the fresh safener-induction MS medium containing 30 mg safener (R29148)/l every 2 weeks. Several normal shoots appeared from five of 12 (41.7%) *ipt*-shooty clones within 3 months of safener induction. These shoots were transferred to the same medium, grew normally and rooted.

4. DNA analysis of normal plants. Normal shoots that appeared from *ipt*-shooty clones were subjected to PCR analysis. A predicted GUS fragment and an excision fragment were amplified, but not a predicted *ipt* fragment. We examined the copy number of normal shoots and *ipt*-shooty clones by Southern analysis. All of the normal shoots contained only a single copy of transgene. These results indicate that the *ipt*-type MAT vector can increase both the regeneration efficiency of transgenic plants and the generation efficiency of marker-free transgenic plants in combination with the *iaa*M/H genes.

7.3.4 Transgene Stacking

Presently, most commercial transgenic crops are altered with respect to single agronomic traits. The next step is to introduce many genes and manipulate complex traits of crops. When the desired traits are additionally introduced to transgenic plants through re-transformation, the presence of marker genes in transgenic plants precludes the use of the same marker gene for the selection of double-transformed plants. While a large number of desired traits and genes are worth incorporating into crops, there are only a limited number of suitable marker genes. Therefore, a transformation system for removing a selection marker is a prerequisite for stacking multiple genes by re-transformation.

We demonstrated that the MAT vector could efficiently generate marker-free transgenic plants. However, since the MAT vector uses the *R/RS* system for removal of a selectable marker gene, one recognition site (RS) remains in the genome of marker-free transgenic plants. In yeast, the *R/RS* system can mediate recombination between two RS sites, which are present about 180 kb apart on one chromosome or on two non-homologous chromosomes (Matsuzaki et al. 1990). Recombination leads to chromosomal excision, inversion or translocation. When many genes of interest are introduced through re-transformation using MAT vectors, recombination between the remaining RS sites and the introduced RS sites might cause chromosomal rearrangement. Therefore, we re-transformed transgenic plants using the MAT vector and examined their chromosomal DNA by Southern blot analysis.

We inserted the chimeric GFP gene with a Nos promoter into the lacZ' multi-cloning sites of the GST-MAT vector pMAT8 as a model gene of interest due to construct pMAT8:NosGFP plasmid (Fig. 7.2B). We used a marker-free transgenic tobacco line, the 132BMO6 line produced via transformation with pNPI132, for re-transformation with pMAT8:NosGFP (Sugita et al. 2000b). The transformation process of tobacco using pMAT8:NosGFP is as follows:

1. Gene transfer by infection. Twenty pieces of leaf segments of transgenic tobacco (132BM06) plants were infected with *A. tumefaciens* containing the pMAT8:NosGFP vector and co-cultivated on hormone-free MS agar medium containing 50 mg acetosyringone/l for 3 days.
2. Regeneration of transgenic plants. The explants were transferred to hormone-free MS agar medium containing 500 mg carbenicillin/l but not kanamycin (nonselective medium). One month after infection, 36 regenerated adventitious buds were separated from the leaf segments and transferred to the same medium. After 1 month of cultivation, we visually identified 34 (94%) *ipt*-shooty phenotypes and subjected them to PCR analysis.
3. Appearance of marker-free plants. Twenty (59%) *ipt*-shooty explants, in which the predicted excision fragments were amplified by PCR, were subcultured to fresh safener-induction MS medium containing 30 mg safener (R29148)/l every month. Several normal shoots appeared from 7 of 20 (35%) *ipt*-shooty clones within 3 months of safener induction. These shoots were transferred to the same medium, grew normally and rooted.
4. DNA analysis of normal plants. Seven normal shoots that appeared from independent *ipt*-shooty clones were subjected to PCR analysis. Predicted *npt*II and *gus*A fragments, and an excision fragment of pNPI132 were amplified in all seven normal plants. In five of seven normal plants, a predicted GFP fragment and an excision fragment of pMAT8:NosGFP were amplified, as was a predicted *ipt* fragment in one of five GFP-positive normal plants. These results indicated that four marker-free transgenic plants in which multiple transgenes were stacked had been generated from 20 excision-positive *ipt*-shooty lines (20%). We investigated the integrated *npt*II and *gus*A genes of four marker-free transgenic plants by Southern blot analysis. Genomic DNA of the untransformed 132BMO6 line and four marker-free transgenic plants were digested with two restriction enzymes and hybridized with the GUS coding regions. No rearrangement of the integrated genes was detected (Fig. 7.8). These results showed that the GST-MAT vector did not cause DNA rearrangement between the first and second transformations. Since the GST-MAT vector tends to generate marker-free transgenic plants containing a single transgene, this vector might be effective for avoiding undesirable DNA rearrangements.

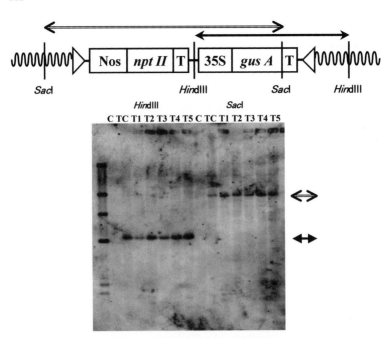

Fig. 7.8. Southern analysis of transgene-stacked plants. Genomic DNA were digested with two restriction enzymes (*Hind*III or *Sac*I) and hybridized with the GUS coding regions. *Lanes C* Non-transgenic control, *TC* original marker-free transgenic tobacco plants, *T1–T5* independent transgene-stacked marker-free transgenic tobacco plants

7.4 Single-Step Transformation

In the *ipt*-type MAT vector, the *ipt* gene is used as a selectable marker gene to regenerate transgenic plants and select marker-free transgenic plants. However, monocot plant species are known to regenerate through somatic embryogenesis rather than cytokinin-dependent organogenesis. There has been no report that the *ipt* gene can induce the regeneration of shoots and cause the *ipt*-shooty phenotype in monocot plants. To study the effect of the *ipt* gene in monocot plants, we introduced the 35S-*ipt* gene into rice plants using a conventional transformation method and investigated the phenotypic characteristics of transgenic plants containing the *ipt* gene. Transgenic rice plants containing the *ipt* gene showed the *ipt*-shooty phenotype, which lacks apical dominance and rooting ability. These results indicate that the *ipt* gene can be used as a selectable marker in rice.

We constructed the MAT vector pNPI30GFP, in which the *ipt*, *R* and GFP genes with the 35S promoters are combined with the site-specific recombination *R*/*RS* system for removal from transgenic cells after transformation. The *gusA* gene and the *hpt* gene outside of the "hit and run" cassette of the MAT vector are used as model genes of interest (Fig. 7.2C). To examine the effect of pre-culture periods on the regeneration of transgenic shoots, we used scutellum tissues that had been excised from seeds pre-cultured for different periods for transformation.

Rice seeds (*Oryza sativa* L. cv. Kitaake) were sterilized and cultured on ESS1 (N6 salts, MS vitamins, 30 g sucrose /l, 30 g maltose/l, 0.4% gelrite) or N6Cl medium containing 2 mg2.4-D/l for 5, 8 and 12 days. Scutellum tissues were excised from these pre-cultured rice seeds and infected with EHA105 containing pNPI130GFP. After 3 days of co-culture on co-ESS1 or co-N6Cl medium containing 10 mg acetosyringone/l, the scutellum tissues were transferred to hormone-free MSR medium containing 500 mg carbenicillin/l. After 4 weeks, we obtained regenerated shoots from 29 of 85 (34.1%) scutellum tissues that had been pre-cultured for 5 days, 40 of 82 (48.8%) for 8 days and 60 of 106 (56.6%) for 12 days on ESS1 medium, and from 14 of 68 (20.8%) for 5 days, 28 of 55 (50.9%) for 8 days and 23 of 38 (60.5%) for 12 days on N6Cl medium (Fig. 7.9). We analyzed regenerated shoots by PCR and confirmed the presence of the *gus* gene and excision of the *ipt* gene. We found that 17 of 24 (70.8%) regenerated shoots with 5 days of pre-culture were transgenic plants, 6 of 38 (15%) for 8 days and 4 of 32 (1.3%) for 12 days on ESS1 medium, and 7 of 14 (50%) for 5 days, 8 of 20 (40%) for 8 days and 2 of 18 (11.1%) for 12 days on N6Cl medium. These results show that, as the pre-culture period decreases from 12 to 5 days, the ratio of transgenic shoots to non-transgenic shoots increases. PCR analysis also indicated that all of the transgenic plants with 5 and 8 days of pre-culture were marker-free while 50.0% of those with 12 days of pre-culture on ESS1 medium were marker-free. Marker-free transgenic rice plants were regenerated directly from the infected scutellum tissues without the production of *ipt*-shooty intermediates. These results show that the pre-culture period of rice seed is an important factor for the generation of marker-free transgenic rice plants using the *ipt*-type MATVS.

In our two-step transformation method using the *ipt*-type MAT vector, we selected transgenic plants as *ipt*-shooty phenotypes and then induced the regeneration of marker-free transgenic plants. We need more sub-cultures to visually discriminate transgenic shoots from non-transgenic escapes, and to extend normal phenotypic marker-free transgenic shoots from *ipt*-shooty clones. The single-step transformation method allows us to directly regenerate marker-free transgenic rice plants from infected scutellum tissues (Fig. 7.4). In the current transformation method using hygromycin, it takes about 3 months to generate transgenic rice plants and 3 months more to segregate marker-free transgenic progenies by crossing. Our method can generate marker-free transgenic rice plants at high frequency within 1 month. Therefore, MATVS can save both time and effort in generating marker-free transgenic rice plants.

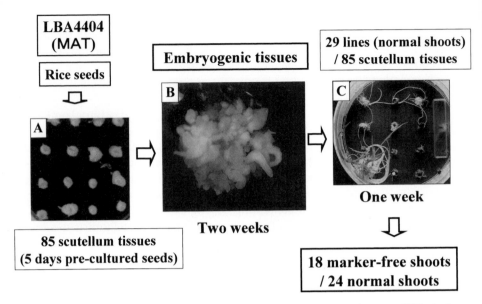

Fig. 7.9A–C. Transformation of rice using the *ipt*-type MAT vector (pNPI130GFP). **A** Co-cultivation of rice scutellum tissues, **B** induction of rice embryogenic tissues, **C** regeneration of marker-free transgenic rice plant (*MF*)

7.5 Cloning Vector for Desired Genes

The binary vector plasmid pBI121 was reconstructed to produce the new vector pTL7 which contains only lacZ' multi-cloning sites and an *Sse*I site between the left and right border sequences. The *Pst*I, *Sse*I and *Sph*I sites of the multi-cloning sites were deleted in pTL7. A gene of interest which lacks an *Sse*I site can be inserted into some sites on the lacZ' multi-cloning sites of pTL7 by blue/white colony selection. The MAT cassettes can be inserted into the *Sse*I site of pTL7 containing the gene of interest (Fig. 7.2D).

We improved the binary vector plasmid pTL7 to clone a gene of interest with an *Sse*I site. We digested the plasmid pDONR201 (Life Technology) with the restriction enzymes *Hpa*I and *Nru*I. DNA fragments including a *ccd*B gene flanked by the attP1 and attP2 sites were ligated to the *Sma*I site of pTL7 to create pTLGW. The MAT cassette can be inserted into the *Sse*I site of pTLGW and the gene of interest can be cloned using the Gate Way Cloning System (Fig. 7.2D).

7.6 Concluding Remarks

Cytokinins are major plant growth regulators that control growth and development in plants. Skoog and Miller (1957) demonstrated that the addition of cytokinin to culture medium containing auxin could induce cell division in cultured tobacco tissues. Later, it was established that cytokinin, in combination with auxin, stimulated plant cell division and determined the direction of plant cell differentiation. A high cytokinin-to-auxin ratio triggers shoot formation but inhibits root induction, whereas a low ratio produces the opposite effect. Plant growth regulators (cytokinin, auxin) are widely used to control differentiation in cultured plant cells and tissues. On the other hand, *ipt*-type MAT vectors use the *ipt* gene to manipulate endogenous hormone levels instead of having to apply them exogenously. However, the endogenous levels of plant hormones and the cell responses to plant growth regulators are very different depending on the plant species, plant tissue and developmental stages. Therefore, the state of plant materials, the choice of a promoter of the *ipt* gene and the tissue culture conditions greatly affect the generation efficiency of marker-free transgenic plants and need to be independently evaluated for each plant species.

The *ipt* gene fused with a 35S promoter has been reported to increase the cytokinin content in transgenic tobacco cells (Smigocki and Owens 1988). Considerable variation of the cytokinin content was observed and elevated cytokinin levels were correlated with a more rapid induction of shoots. However, the highest cytokinin concentration was observed in unorganized transgenic tissues. Unfavorably high cytokinin levels inhibited shoot induction and proliferated undifferentiated cells. These results suggest alternative routes for the regeneration of transgenic plants using the *ipt* gene. (1) A promoter of the *ipt* gene is changed to regulate its expression and optimize the cytokinin level to regenerate transgenic shoots. (2) The *ipt* gene is fused with a highly expressed promoter to overproduce cytokinin. Unfavorably high cytokinin levels lead to the proliferation of undifferentiated transgenic cells. Cytokinin that has leaked out of transgenic cells causes the regeneration of marker-free transgenic cells that have lost the *ipt* gene.

In this chapter, we have described two strategies for the generation of marker-free transgenic plants using *ipt*-type MAT vectors:

1. Two-step transformation. First, the *ipt* gene regenerates transgenic shoots and then the *R/Rs* system removes it to generate marker-free transgenic shoots. We changed a promoter of the *ipt* and *R* genes and regulated their expression to optimize the cytokinin level for the regeneration of transgenic shoots and to induce excision events after selection.
2. Single-step transformation. First, the *ipt* gene proliferates transgenic calluses and then the *R/Rs* system removes it to regenerate marker-free transgenic shoots. The overproduction of cytokinin by the highly expressed 35S-*ipt* gene leads to the proliferation of transgenic calluses rather than

regenerated transgenic shoots. Recombinase produced by the 35S-*R* gene causes excision events of the *ipt* gene during the proliferation of transgenic calluses. Changes in cytokinin levels trigger shoot induction from transgenic cells that have lost the *ipt* gene.

With regard to two-step transformation, we need more sub-cultures to visually discriminate transgenic shoots from non-transgenic escapes, and to extend normal phenotypic marker-free transgenic shoots from *ipt*-shooty clones. Since single-step transformation does not require a step for the selection of *ipt*-shooty intermediates, it may be an effective shortcut for the generation of marker-free transgenic plants.

References

Akiyoshi DE, Klee H, Amasino RM, Nester EW, Gordon MP (1984) T-DNA of *Agrobacterium tumefaciens* encodes an enzyme of cytokinin biosynthesis. Proc Natl Acad Sci USA 81:5994–5998

Barry GF, Rogers SG, Fraley RT, Brand L (1984) Identification of a cloned cytokinin biosynthetic gene. Proc Natl Acad Sci USA 81:4776–4780

Belzile F, Lassner MW, Tong Y, Khush R, Yoder JI (1989) Sexual transmission of transposed *Activator* elements in transgenic tomatoes. Genetics 123:181–189

Ebinuma H, Komanine A (2001) MAT (multi-auto-transformation) vector system. The oncogenes of *Agrobacterium* as positive markers for regeneration and selection of marker-free transgenic plants. In Vitro Cell Dev Biol Plant 37:114–119

Ebinuma H, Sugita K, Matsunaga E, Yamakado M (1997a) Selection of marker-free transgenic plants using the isopentenyl transferase gene as a selectable marker. Proc Natl Acad Sci USA 94:2117–2121

Ebinuma H, Sugita K, Matsunaga E, Yamakado M, Komamine A (1997b) Principle of MAT vector. Plant Biotechnol 14:133–139

Ebinuma H, Sugita K, Matsunaga E, Endo S, Kasahara T (2000) Selection of marker-free transgenic plants using the oncogenes (*ipt, rol A, B, C*) of *Agrobacterium* as selectable markers. In: Jain SM, Minocha SC (eds) Molecular biology of woody plants, vol 2. Kluwer, Dordrecht, pp 25–46

Ebinuma H, Sugita K, Matsunaga E, Endo S, Yamada K, Komamine A (2001) Systems for the removal of a selection marker and their combination with a positive marker. Plant Cell Rep 20:383–392

Endo S, Kasahara T, Sugita K, Matsunaga E, Ebinuma H (2001) The isopentenyl transferase gene is effective as a selectable marker gene for plant transformation in tobacco (*Nicotiana tabacum* cv. Petite Havana SR1). Plant Cell Rep 20:60–66

Estruch JJ, Prinsen E, Onckelen HV, Shell J, Spena A (1991) Viviparous leaves produced by somatic activation of an inactive cytokinin-synthesizing gene. Science 254:1364–1367

Fedoroff N (1989) Maize transposable elements. In: Berg DE, Howe MM (eds) Mobile DNA. Am Soc Microbiol, Washington, pp 375–411

Holt DC, Lay VJ, Clarke ED, Dinsmore A, Jepson I, Bright SWJ, Greenland AJ (1995) Characterization of the safener-induced glutathione S- transferase isoform II from maize. Planta 196: 295–302

James DJ, Passey AJ, Barbara DJ (1990) *Agrobacterium*-mediated transformation of the cultivated strawberry (*Fragaria _ Anannassa* Duch.) using disarmed binary vectors. Plant Sci 69:79–94

Li Y, Hagen G, Guilfoyle TJ (1992) Altered morphology in transgenic tobacco plants that overproduce cytokinins in specific tissues and organs. Dev Biol 153:386–395

Matsuzaki H, Nakajima R, Nishiyama J, Araki H, Oshima Y (1990) Chromosome engineering in *Saccharomyces cerevisiae* by using a site-specific recombination system of a yeast plasmid. J Bacteriol 172:610–618

Medford JI, Horgan R, El-Sawi Z, Klee HJ (1989) Alterations of endogenous cytokinins in transgenic plants using a chimeric isopentenyl transferase gene. Plant Cell 1:403–413

Onouchi H, Yokoi K, Machida C, Matsuzaki H, Oshima Y, Matsuoka K, Nakamura K, Machida Y (1991) Operation of an efficient site-specific recombination system of *Zygosaccharomyces rouxii* in tobacco cells. Nucleic Acids Res 19:6373–6378

Ooms G, Kaup A, Roberts J (1983) From tumour to tuber; tumour cell characteristics and chromosome numbers of crown gall-derived tetraploid potato plants (*Solanum tuberosum* cv. "Maris Bard"). Theor Appl Genet 66:169–172

Schmulling T, Beinsberger J, Greef JD, Schell J, Onckelen HV, Spena A (1989) Construction of a heat-inducible chimeric gene to increase the cytokinin content in transgenic plant tissue. FEBS Lett 249:401–406

Schwartzenberg KV, Doumas P, Jouanin L, Pilate G (1994) Enhancement of the endogenous cytokinin concentration in poplar by transformation with *Agrobacterium* T-DNA gene *ipt*. Tree Physiol 14:27–35

Skoog F, Miller CO (1957) Chemical regulation of growth and organ formation in plant tissues cultured in vitro. Symp Soc Exp Biol 11:118–130

Smart MC, Scofield SR, Bevan MW, Dyer TA (1991) Delayed leaf senescence in tobacco plants transformed with *tmr*, a gene for cytokinin production in *Agrobacterium*. Plant Cell 3:647–656

Smigocki AC, Hammerschlag FA (1991) Regeneration of plants from peach embryo cells infected with a shooty mutant strain of *Agrobacterium*. J Am Soc Hortic Sci 116:1092–1097

Smigocki AC, Owens LD (1988) Cytokinin gene fused with a strong promoter enhances shoot organogenesis and zeatin levels in transformed plant cells. Proc Natl Acad Sci USA 85: 5131–5135

Smigocki AC, Owens LD (1989) Cytokinin-to-auxin ratios and morphology of shoots and tissues transformed by a chimeric isopentenyl transferase gene. Plant Physiol 91:808–811

Sugita M, Gruissem W (1987) Developmental, organ-specific, and light-dependent expression of the tomato ribulose-1,5-bisphosphate carboxylase small subunit gene family. Proc Natl Acad Sci USA 84:7104–7108

Sugita K, Matsunaga E, Ebinuma H (1999) Effective selection system for generating marker-free transgenic plants independent of sexual crossing. Plant Cell Rep 18:941–947

Sugita K, Kasahara T, Matsunaga E, Ebinuma H (2000a) A transformation vector for the production of marker-free transgenic plants containing a single copy transgene at high frequency. Plant J 22:461–469

Sugita K, Matsunaga E, Kasahara T, Ebinuma H (2000b) Transgene stacking in plants in the absence of sexual crossing. Mol Breed 6:529–536

Thomashow LS, Reeves S, Thomashow MF (1984) Crown gall oncogenesis: evidence that a T-DNA gene from the *Agrobacterium* Ti plasmid pTiA6 encodes an enzyme that catalyses synthesis of indoleacetic acid. Proc Natl Acad Sci USA 81:5071–5075

Vahala T, Eriksson T, Tillberg E, Nicander B (1993) Expression of a cytokinin synthesis gene from *Agrobacterium tumefaciens* T-DNA in basket willow (*Salix viminalis*). Physiol Plant 88:439–445

Wabiko H, Kagaya M, Kodama I, Masuda K, Kodama Y, Yamamoto H, Shibano Y, Sano H (1989) Isolation and characterization of diverse nopaline type Ti plasmids of *Agrobacterium tumefaciens* from Japan. Arch Microbiol 152:119–124

8 Safety Assessment of Insect-Protected Crops: Testing the Feeding Value of *Bt* Corn and Cotton Varieties in Poultry, Swine and Cattle

B. HAMMOND, E. STANISIEWSKI, R. FUCHS, J. ASTWOOD, and G. HARTNELL

8.1 Introduction

Health and safety is the number one priority in developing new food products, including those developed through biotechnology. Before any food crop produced using modern biotechnology can be sold, it must undergo rigorous testing and assessments, spanning multiple years and multiple approaches, to establish its safety as a food product. No other food crops in history – including foods currently available on grocers' shelves – have been tested and regulated as thoroughly as foods developed through biotechnology.

The practical safety of biotechnology-derived foods is assured by meeting the high standards established for the safety of all foods and conducting rigorous safety assessments based on the latest guidance from regulatory agencies and national and international scientific organizations, using the most robust scientific methods.

8.1.1 Food Safety Standards

National and international regulatory authorities require that food produced through biotechnology must meet the same safety standards as food grown conventionally; that is, there must be "reasonable certainty that no harm will result from intended uses under the anticipated conditions of consumption". The food safety standard for biotechnology-derived food therefore is that these foods must be *as safe as* food produced by conventional varieties.

The World Health Organization (WHO) and Organization for Economic Cooperation and Development (OECD) have established the safety assessment process called "substantial equivalence" to assure that new foods are indeed as safe as food produced from conventionally bred crops. This process considers two main categories of risk: the properties of the introduced trait, and any effects created by the introduction or expression of the new trait in the crop or food. This is a comparative safety assessment: conventional foods serve as a reference point for all safety testing.

Molecular Methods of Plant Analysis, Vol. 22
Testing for Genetic Manipulation in Plants
Edited by J.F. Jackson, H.F. Linskens, and R.B. Inman
© Springer-Verlag Berlin Heidelberg 2002

8.1.2 Testing for Food and Feed Safety

Data are systematically collected to assess safety. Five main categories of testing may be described as:

- Safety of the new trait, most often the "introduced protein"
- Comparison of the agronomic characteristics of the new plant to traditionally bred plants, also called "agronomic equivalence"
- Comparison of the nutritional and biochemical composition of the new food to traditional food, also called "compositional equivalence"
- Safety of the resulting food or feed established by comparative toxicology testing
- Nutritional wholesomeness of the new food or feed established by testing in farm animals

Safety of New Trait (Introduced Protein). The unique characteristic of biotechnology that is most relevant to strategies to assess safety is that most traits result from the introduction of one or a few proteins. This places biotechnology at considerable advantage relative to the safety assessment of the products produced by other technologies such as traditional pesticides or pharmaceuticals, since there are very limited ways in which proteins can be harmful. Indeed most proteins are desirable – people and animals are evolved to "eat" proteins, an essential nutritional requirement in the diet. In contrast, small organic molecules which are intended to either affect the physiological state of a person (i.e. a pharmaceutical) or kill things (i.e. a pesticide) require answers to a set of possibilities much more diverse than for proteins, and therefore require a different set of safety tests.

The introduced protein is extensively characterized to understand how it functions and how similar it is to proteins already present in foods. For example, the protein used to confer tolerance to Monsanto's Roundup herbicide, is a member of a family of proteins that already is present in most foods, has a safe function and may be generally regarded as safe due to its ubiquity in the food supply. The amount of the introduced protein is measured in key raw agricultural commodities to evaluate consumption patterns. The amino acid sequence of the introduced protein(s) is compared to known toxins and allergens to assure the protein is neither a toxin or allergen or closely related to either. As mentioned above, since proteins are a key component in food, the digestibility of the protein can play an important role in predicting safety. Clearly, a protein that is stable to digestion is more likely to cause harm relative to a protein which is rapidly digested. To confirm that a protein is unlikely to cause harm, it is tested for toxicity by testing in animals at high levels (thousands to hundreds of thousands times greater than the highest predicted consumption) to assure no adverse effects. Not surprisingly, given the nature and digestibility of proteins, no toxicities have been observed in these tests – the introduced proteins are considered "practically non-toxic."

The likelihood of the protein being an established allergen or becoming an allergen is also assessed in detail, according to international standards. Each of the proteins that Monsanto used in commercial crops has been shown to share no significant similarities to the properties of allergens and hence pose no significant concerns from an allergenic perspective.

Agronomic Equivalence. As part of the overall safety assessment of a crop developed via biotechnology, various agronomic and morphological parameters of the crop are compared with those of the conventional counterpart to assure that there are no meaningful unintended agronomic or morphological changes caused by the transformation process or the introduced genes/trait. The morphology, yield and other agronomic parameters are sensitive indicators of changes in the metabolism or physiology of the plant. Plants developed through biotechnology must meet very stringent criteria. Some of the parameters assessed for corn products include: plant height, time to pollination, ear length, leaf shape and color, susceptibility to disease, drought and yield. This is a survey for "unintended effects" which, as is done in conventional breeding practices, helps eliminate plants with unintended effects.

Compositional Equivalence. A key focus of substantial equivalence is a comprehensive comparison of key nutrients, anti-nutrients, toxins and other compounds naturally found in the foods. Biotechnology-derived and conventional plant varieties are grown under a variety of field conditions to assess the composition under commercially representative growing conditions. The key macro-nutrients (e.g. protein, oil, carbohydrate, fiber, ash and moisture), as well as the levels of the individual amino acids, fatty acids, vitamins and minerals, are assessed as are the levels of key toxicants, anti-nutrients and allergens. The values for the biotechnology-derived crop are compared with those of both the parental control and other commercial varieties of that crop to assess whether the range of values obtained for the biotechnology-derived crop fall within the levels of the conventional varieties, as well as within the published values for that crop. Typically 60–90 different components are analyzed. For Roundup Ready soybean, Monsanto conducted over 1,800 compositional analyses.

Whole Food Comparative Toxicology Testing. To confirm that new foods and feeds developed by biotechnology are as safe as traditional foods or feeds, subchronic (28- or 90-day) comparative toxicity studies are performed with grain from both the biotechnology-derived and conventional plant varieties. This is a very robust and internationally recognized testing approach that assesses the safety of the intentionally introduced genes/proteins as well as any unintended consequences due to insertion of the genes into the plant genome or due to other unintended consequences relative to conventional plant varieties.

Nutritional Wholesomeness in Farm Animal Testing. Monsanto conducted animal-feed performance studies on biotechnology-derived crops in farm animals such as dairy cows, beef cattle, swine and poultry, to assess if the grain from these crops is nutritionally equivalent to feed from conventional crops. These studies can also detect unintended consequences relative to conventional plant varieties and provide confirmation of: (1) the safety of the introduced trait (protein), (2) the nutritional/compositional equivalence and (3) the safety of the whole food as determined in toxicology testing. Feeding studies in farm animals with insect-protected corn will be the subject of this chapter.

8.2 Insect Protection Traits

Biotechnology has made it possible to develop plant varieties with improved qualities such as self-protection against insect pests. Reducing injury caused by pests improves the health of plants. Healthy plants produce safe and nutritious foods benefiting consumers. Self-protected plants require less chemical insecticide application, reducing exposure to farm applicators and the environment. Reduction in the use of non-selective chemical insecticides may also increase beneficial organisms in the field. This chapter reviews the safety and benefits of insect-protected crop varieties that have been developed via biotechnology and introduced into commerce during the last few years. This assessment includes information on the safety of the introduced insect-protection traits. Extensive agronomic and compositional studies have shown insect-protected plants to be "substantially equivalent" to traditional crop varieties. Data from recently completed farm-animal feeding studies with insect-protected corn and cotton will be summarized, providing additional confirmation of the safety of these crop varieties.

Insect-protected plants commercialized to date are genetically enhanced to produce insect-control proteins *in planta* like those made by *Bacillus thuringiensis* (*Bt*) (Fischhoff et al.1987; Vaeck et al.1987; Perlak et al.1990). *Bt* is a ubiquitous gram-positive soil bacterium that forms crystalline protein inclusions during sporulation (Hofte and Whitely 1989). The inclusion bodies consist of Cry proteins (Cry is an acronym for crystal) which are selectively active against certain lepidopteran, dipteran or coleopteran pests. Microbial *Bt* products containing Cry proteins were first commercialized in 1961 for use in agriculture and have been safely used for nearly 40 years (Baum et al. 1999). The first *Bt* microbial formulations were based on *Bt kurstaki* strain HD1 which produces four Cry proteins active against lepidopteran pests: Cry1Aa, Cry1Ab, Cry1Ac and Cry2Aa. The *cry1Ab* and *cry1Ac* genes in the *Bt* HD1 strain are the prototypes for the genes currently expressed in corn and cotton to provide in planta protection against lepidopteran pests. *In planta* production of these insect-control proteins confers plant protection throughout the growing season. Efficacy of surface-applied microbial *Bt* preparations is limited to a few days since the Cry proteins on the leaf surface are inactivated by sunlight or washed off by rain. Plant tissue levels of Cry proteins required to provide protection against pests is very low, in the part per million range. For example, in YieldGard corn, the levels of Cry1Ab protein range from 9–12 ppm in leaves to 0.3–0.5 ppm in the grain (Sanders et al.1998).

8.3 Benefits

Growers sustain billions of dollars in crop loss or reduced yield due to pests that have the potential to be controlled by Cry proteins (Gianessi and Carpenter 1999). The European corn borer causes stalk damage when second-generation borers enter the corn stalks. Once in the stalk, corn borers are difficult to control with externally applied insecticides.

Corn plants that are stressed by insect damage are more susceptible to fungal infection, particularly under drought conditions. Certain fungal species such as *Fusarium verticillioides* and *Fusarium proliferatum* are found in corn wherever it is grown. These fungi produce secondary metabolites or mycotoxins known as fumonisins. Fungi can be introduced into the corn plant by corn borers feeding on stalk and ear tissue. The resulting damage to developing grain enables spores of the toxin-producing fungi to enter the kernel and proliferate producing corn ear rot. Fumonisins can make the grain unhealthy for human or animal consumption. At levels greater than 10–20 ppm in grain, fumonisin contamination can cause death or morbidity in horses and swine (Norred 1993). Epidemiological studies have linked consumption of corn containing high levels of fumonisins with an elevated incidence of esophageal and liver cancer in African subsistence farmers (Marasas et al.1988). Fumonisins are not genotoxic, but have been shown to cause liver and kidney cancer in rodents. As a consequence of growing concerns about the potential health implications of fumonisins, and their ubiquitous presence in corn grown around the world, international health agencies (FAO/WHO JECFA etc.) have established a provisional maximum tolerable daily intake (PMTDI) of 2 μg fumonisin per kg body weight dietary exposure for humans (FAO/WHO 2001). Action levels for fumonisin contamination in grain have been set by two countries, Switzerland (1 ppm) and the United States (2–4 ppm for human food, 5 ppm for horses, 20 ppm for swine, and higher levels for poultry and ruminants). For horses and swine, the contaminated corn cannot exceed 20% of the diet mix so fumonisin exposure to animals is further diluted.

Since insect damage caused by corn borers can predispose plants to fungal growth and fumonisin contamination, protection against corn borers can reduce fungal and mycotoxin contamination. Munkvold et al. (1999) were the first to show that *Fusarium* ear rot and fumonisin contamination were dramatically reduced in insect-protected *Bt* corn compared with non-*Bt* corn over several years of field trials. This was confirmed in additional field trials carried out with insect-protected *Bt* corn grown over 3 years in Illinois (Dowd 2000) as well as field trials carried out in France, Spain and Italy (Cahagnier and Melcion 2000; Pietri and Piva 2000). All of these studies showed a reduction in fumonisin levels when compared with non-*Bt* corn. Protection of corn plants against corn borer damage can prevent plant injury, reducing the potential for fungal and fumonisin contamination, which may improve the safety of corn grain for human and animal consumption.

In planta protection against insect pests can reduce the use of insecticides on the plant. Cotton plants used to be heavily sprayed with insecticides in the south and southwestern United States to control tobacco budworm and cotton and pink bollworm. Following the commercial introduction of insect-protected *Bt* cotton, there has been a significant reduction (millions of pounds annually) in the use of insecticides to control these pests (Gianessi and Carpenter 1999).

8.4 Safety Assessment of the Cry Insect-Control Proteins

The microbial *Bt* formulations that contain the Cry insect-control proteins have an exemplary safety record following 40 years of use in agriculture (Betz et al. 2000). There are currently at least 180 registered microbial *Bt* products (EPA 1998b) and over 120 microbial products in the European Union. Microbial *Bt* formulations have been applied to human food crops to control pests and to drinking water to control mosquito larvae. Microbial *Bt* formulations have been exempted from tolerances for use on agricultural food crops. The WHO International Program on Chemical Safety Environmental Health Criteria Report on *Bt* concluded that: "*Bt* has not been documented to cause any adverse effects on human health when present in drinking water or food" (IPCS 2000).

In the United States, microbial *Bt* formulations containing Cry insect-control proteins are regulated by the Environmental Protection Agency (EPA), which has regulatory jurisdiction over pesticides. EPA also regulates genetically modified plants that produce Cry proteins *in planta* to provide insect protection. Many toxicology studies have been carried out over the years to assess the safety of *Bt* microbial formulations for mammals and other non-target organisms. Based on a recent review of all the completed safety studies, EPA concluded that "Toxicology studies submitted to the U.S. Environmental Protection Agency to support the registration of *B. thuringiensis* subspecies have failed to show any significant adverse effects in body weight gain, clinical observations or upon necropsy." (EPA 1998a; McClintock et al. 1995). In the aforementioned WHO monograph, similar conclusions were made following peer review of published literature on the safety assessment of microbial *Bt* formulations "Owing to their specific mode of action, *Bt* products are unlikely to pose any hazard to humans or other vertebrates . . ." (IPCS 2000). The specific mode of action of the Cry insect control proteins is the basis for their selectivity and lack of toxicity towards non-target organisms.

8.5 Mode of Action

Cry proteins are produced as protoxins that are proteolytically activated upon ingestion (Hofte and Whitely 1989). Cry proteins bind to specific receptors on the surface of midgut cells of susceptible insects and form ion-selective channels in the cell membrane (English and Slatin 1992). The cells swell due to an influx of water which leads to cell lysis, the insect stops eating and dies (Knowles and Ellar 1987).

If receptor binding does not occur, the Cry protein will have no effect on that organism. Results of several studies have failed to find Cry-protein-specific receptors on gut cell membranes of various non-target mammalian species such as mice, rats, monkeys, and humans (Hofmann et al. 1988; Noteborn et al. 1993). This explains why the Cry insect-control proteins are acutely toxic to target insects at µg/kg body weight doses, but are non-toxic to mammals dosed acutely with greater than 1×10^6 µg/kg Cry proteins (Sjoblad et al.1992; McClintock et al.1995; EPA 1998a).

As a condition for registration with EPA, Cry insect-control proteins that will be introduced into plants must be administered acutely at very high dosages to laboratory rodents as part of an overall hazard safety assessment. The dosage levels administered to rodents are generally in the thousands of mg/kg range. To date, no adverse effects have been observed in rodents dosed acutely with Cry insect-control proteins introduced into plants. This provides extremely large margins of safety for humans or farm animals that might consume corn grain containing these Cry proteins. For example, a 600-kg dairy cow would have to eat 4,800,000 kg of YieldGard corn grain containing 0.5 ppm Cry 1Ab protein to equal the 4,000 mg/kg Cry 1Ab protein dose given acutely to mice. The 4,000 mg/kg Cry1Ab dose produced no toxic effects in mice (Sanders et al.1998). A 60-kg human would have to eat 480,000 kg of YieldGard corn grain to achieve the same dose of 4,000 mg/kg Cry1Ab protein given to mice. Based on the absence of mammalian toxicity for Cry1Ab protein and the large margins of safety for dietary exposure, it is concluded that Cry1Ab protein poses no meaningful risk to human or animal health.

Cry proteins have an exemplary safety record and have certain safety advantages when compared to conventional pest control agents. As stated previously, Cry proteins exhibit a high degree of specificity for target insect species and are not toxic to non-target organisms including mammals.

The class of Cry1, Cry2 and Cry3 proteins are readily degraded in vitro using simulated mammalian gastric fluids (Noteborn et al.1994; EPA 1995a,b, 1996, Spencer et al.1996). These proteins are typically 60–130 kDa in size and are degraded in simulated digestion models to polypeptides of less than 2 kDa.

Bioinformatic analyses are used to verify the absence of structural similarity of Cry proteins or their degradation products to known allergens, toxins or pharmacologically active proteins.

The digestive systems of humans and animals effectively degrade proteins in food or feedstuffs into their amino acid components, which are absorbed and used for the synthesis of new protein macromolecular components to support growth, maintenance and reproduction. Thus, Cry proteins would not be appreciably absorbed intact from the gastrointestinal tract to bioaccumulate in fatty tissue as is the case with a halogenated chemical insecticide. As a consequence of their degradability by digestive proteases, no residues of Cry proteins have been detected in tissues of animals fed grain containing these proteins. For example, no intact or immunologically reactive fragments of the Cry1Ab protein were detected in pork loin muscle tissue from grower-finisher pigs fed Bt corn (YieldGard) (Weber et al. 2000). In another example, no Cry protein was detected in the milk of lactating dairy cows fed green chop from insect-protected Bt corn (Faust and Miller 1997). Additional studies which confirm the absence of detection of Cry proteins in edible tissues of farm animals fed insect-protected Bt corn have been recently completed and will be published in the near future.

8.6 Substantial Equivalence Based on Compositional Analysis

Insect-protected Bt corn and cotton crops have been shown to be substantially equivalent (comparable in composition) to their non-Bt counterparts. No biologically meaningful differences in the composition of nutrients/anti-nutrients in grain, seed, oil, silage or other crop byproducts have been observed between Bt-expressing crops and their non-Bt counterparts (Berberich et al. 1996; Sanders et al. 1998).

Analysis of the agronomic and morphological characteristics of insect-protected Bt crops confirm the efficacy and stability of the introduced traits and the lack of significant unintended effects that may be attributable to the genetic modification process. Insect-protected Bt crops meet the stringent product performance standards established for new plant varieties. Evaluations consisting of plant vigor, growth habit characteristics, yield, crop quality, and insect and disease susceptibility have shown that insect-protected Bt crops are morphologically and agronomically equivalent to the non-transgenic plants from which they are derived.

It is beyond the scope of this chapter to review the thousands of compositional analyses and field-test results comparing insect-protected Bt crops to their conventional counterparts. Results of these studies can be found elsewhere (Berberich et. al. 1996; Sanders et. al. 1998).

Detailed molecular analyses have been performed on each insect-protected Bt crop to characterize and confirm that the intended genetic material has been introduced. Further analyses confirm that the Cry proteins are produced *in planta* as predicted from the molecular characterization.

8.7 Current Products

Today, insect-protected *Bt* cotton and corn developed by Monsanto have been commercialized in the United States and one or more of these products are marketed in Argentina, Australia, Canada, China, France, Portugal, Romania, South Africa, and Spain (James 1998, 1999). These plants express one of several Cry proteins for the control of lepidopteran insect pests.

8.8 Grower Acceptance

During the 5 years since commercial introduction, US growers have rapidly adopted insect-protected *Bt* crops as an effective tool to enhance high-yield sustainable agriculture. Total planted acreage in the U.S. for insect-protected *Bt* cotton, and corn exceeded 16 million acres in 1998 (Gianessi and Carpenter 1999), comprising 17 and 18% of the total corn and cotton acreage, respectively. According to reports by Clive James (1997, 1998, 1999), the global number of acres of insect-protected *Bt* plants has increased from approximately 10 million in 1997 to 20 million in 1998 and 29 million in 1999. The benefits of decreased pest management costs, increased yields and greater crop-production flexibility are responsible for the rapid adoption of these crops (Culpepper and York 1998; Marra et al.1998). The Economic Research Service of the US Department of Agriculture reports that the use of certain insect-protected *Bt* crops is associated with "significantly higher yields" and "fewer insecticide treatments for target pests" (Klotz-Ingram et al. 1999).

8.9 Future Products

The next generation of insect-protected *Bt* plants will contain multiple *cry* genes, thereby providing growers with a product that offers a broader spectrum of pest control and reduced potential for development of insect resistance. With more than 100 *cry* genes described (Crickmore et al. 1998) and dozens of plants transformed to produce Cry proteins, there is significant potential for expanding the role of *Bt*-mediated plant protection. For example, a new insect-protected *Bt* corn variety has been developed that provides protection against corn rootworm, a pest that causes significant damage to the roots of corn grown in the midwestern and eastern United States. Since most of the insecticide applied to corn is intended to control the corn rootworm pest, the introduction of this new variety will result in a significant reduction in the amount of insecticide applied to corn plants. These new crop varieties will be subjected to the same rigorous safety assessment as current biotechnology-derived products in the marketplace to confirm that they can be used safely for both human and animal food/feed.

Although evidence presented demonstrates the safety of insect-protected *Bt* crops such as corn and cotton, developers of these crop varieties have undertaken a variety of farm-animal feeding studies to support customer and consumer acceptance of this technology. These studies were not designed as toxicity studies, but are intended to compare responses of animals fed insect-protected *Bt* crops with animals fed conventional counterparts under commercial conditions of use.

8.10 Farm-Animal Studies

As the adoption of biotechnology-derived crops grew between 1996 and 2000, the animal production industry and related associations began receiving inquiries regarding the performance of farm animals fed biotechnology-derived crops. The primary issue is to affirm that palatability is unchanged and that farm animals fed insect-protected *Bt* crops perform as well as animals fed conventional counterparts of the same crop. Numerous studies with both beef and dairy cattle, broiler and layer chickens, swine and sheep have shown there is no difference in performance in animals fed biotechnology-derived crops compared with animals fed conventional crops (Clark and Ipharraguerre 2001). The following is a summary of recently published farm-animal feeding studies with insect-protected *Bt* crops. An itemized list of studies is presented in Table 8.1.

8.10.1 *Bt* Corn

8.10.1.1 Poultry

One of the first short-term exposure studies utilized insect-protected *Bt* or non-*Bt* corn grain (50% diet wt./wt.) fed to laying hens (6 birds/group) for 5 days. No differences were found in nutrient composition, body weight, digestible organic matter and protein as well as metabolizable energy (Aulrich et al. 1998).

In a longer-term, 38-day broiler study comparing insect-protected *Bt* and non-*Bt* corn, no differences were found in mortality, body weight or feed intake, although there was an improvement ($p < 0.05$) in feed conversion associated with the *Bt* group. Carcass data were not different between groups with

Crop event	Crop	Cry Protein	Animal species (duration test)	Parameters measured	References
MON 810 (YieldGard) and MON 810 × Roundup Ready (Monsanto)	Corn	Cry1Ab	Poultry (42 days)	Weight gain, feed efficiency carcass quality	Taylor et al. (2001)
MON 810 YieldGard (Monsanto)	Corn	Cry1Ab	Poultry (28 days)	Body weight gain, feed efficiency, metabolizable energy, amino acid digestability	Mireles et al. (2000)
MON 810 YieldGard (Monsanto)	Corn	CryAb	Poultry (14 days)	Body weight gain, feed consumption, feed efficiency, metabolizable energy	Gaines et al. (2001a)
MON 810 YieldGard (Monsanto)	Corn	Cry1Ab	Poultry (42 days)	Body weight gain, feed efficiency	Piva et al. (2001b)
MON 810 YieldGard (Monsanto)	Corn	Cry1Ab	Swine (35 days)	Body weight gain, feed efficiency	Piva et al. (2001a)
MON 810 YieldGard (Monsanto)	Corn	Cry1Ab	Swine (5 days)	Digestible energy coefficients	Gaines et al. (2001b)
MON 810 YieldGard (Monsanto)	Corn	Cry1Ab	Swine (100 days)	Body weight gain, feed intake, feed efficiency, carcass yield	Weber et al. (2000)
MON 810 YieldGard (Monsanto)	Corn	Cry1Ab	Beef calves (200 days, repeat for 2 years)	Body weight gain, dry matter intake, feed efficiency	Hendrix et al. (2000)
BollGard II (Cry1Ac and Cry2Ab)	Cotton	Cry1Ac	Dairy cattle (4 weeks)	Dry matter intake, milk yield and composition	Castillo et al. (2001a)
BollGard (Cry1Ac) × RR (Monsanto)	Cotton	Cry1Ac	Dairy cattle (4 weeks)	Dry matter intake, milk yield and composition	Castillo et al. (2001b)
BollGard I (Cry1Ac) (Monsanto)	Corn	Cry2Ab Cry1Ab	Poultry (38 days)	Weight gain, feed efficiency, carcass composition	Brake and Vlachos (1998)
Bt 176 (Syngenta)	Corn	Cry1Ab	Poultry (5 days)	Body weight, digestible organic matter, metabolizable energy	Aulrich et al. (1998)
Bt176 (Syngenta)	Corn	Cry1Ab	Poultry (35 days)	Body weight gain, feed intake, feed efficiency	Halle et al. (1998)
Bt176 (Syngenta)	Corn	Cry1Ab	Dairy cows (13 weeks)	Body weight, milk composition	Barriere et al. (2001)
Bt176 (Syngenta)	Corn	Cry1Ab	Dairy cows (5 weeks)	Diet intake, milk production and composition	Mayer and Rutzmoser (1999)
Bt 11 (Syngenta)	Corn	Cry1Ab	Dairy cows (3 weeks)	Dry matter intake, milk production and composition, rumen function	Folmer et al. (2000a,b)
Syngenta	Corn	Cry1Ab	Beef cows (70 days)	Daily body weight gain	Faust and Miller (1997)
Syngenta, Pioneer, YieldGard event	Corn	Cry1Ab	Dairy cows (2 weeks) Beef (70–80 days, repeat for second year)	Milk composition Weight gain, body condition score	Russell et al. (2000)
Syngenta	Corn	Cry1Ab	Bull calves (~ 8 months) Sheep	Digestibility organic matter (sheep), body weight gain, feed conversion, hot carcass weight, dressing %, abdominal fat (bull calves)	Daenicke et al. (1999)

the exception of improved breast meat yield ($p < 0.05$) in broilers fed insect-protected Bt corn (Brake and Vlachos 1998).

In another trial, insect-protected Bt or non-Bt corn were fed to broiler chickens for 35 days. There were 12 male chicks per treatment and the diets contained 50% corn. No differences in body weight gain, feed intake, feed conversion or protein digestibility were observed (Halle et al. 1998).

Two additional broiler studies compared nutrient composition and bioavailability of insect-protected Bt with non-Bt corn. The first study measured true metabolizable energy (TME) and amino acid digestibility. Each treatment had nine replicate cockerels. There were no differences for TME or amino acid digestibility in birds between the two corn sources. The second study was designed to measure the performance of broiler chickens fed starter feeds for 7–28 days. Each treatment had six replicate cages of eight broiler chicks each. No differences in weight gain or feed efficiency were observed between birds fed insect-protected Bt and non-Bt corn (Mireles et al. 2000).

Broiler chickens were used to compare performance and processing characteristics of birds fed insect-protected Bt (YieldGard) corn and $Bt \times RR$ (YieldGard $\times$ Roundup Ready) corn compared with their respective non-transgenic control lines (similar background genetics) and four non-transgenic commercial reference hybrids included as references (Taylor et al. 2001). Corn was added to diets at levels of 50% (starter diet) to 60% (finisher diet) and diets were fed for 42 days. There were 50 males and 50 female birds in each group (5 pens/sex/treatment, 10 birds/pen). Performance parameters (body weights, weight gain, feed efficiency) were not statistically different ($p > 0.05$) between the groups. Processing characteristics (live/chill weights, thigh, drum, wing and fat pad weights), composition (% moisture, protein, fat in breast and thigh meat) and carcass yield were not different statistically ($p > 0.05$) between any of the groups.

Male 3-day-old broiler chickens were fed insect-protected Bt or non-Bt (near isogenic) corn in a 14-day growth assay to compare the nutritional value of the corn varieties (Gaines et al. 2001a). Birds were assigned randomly to 50 pens with 5–6 birds/pen. In addition to the insect-protected Bt and non-Bt corn varieties, three commercial non-transgenic hybrids were included as reference controls. Body weight gain, feed intake and feed efficiency were compared between the groups. In a separate experiment, birds (35 pens, 6 birds/pen) were fed only the corn hybrids with a mineral and vitamin supplement. Metabolizable digestability coefficients were determined for each group. Both of these experiments were replicated. It was concluded that the insect-protected Bt corn hybrid was nutritionally equivalent to its near isogenic non-Bt control.

Male broiler chickens were allotted to 6 treatments with 72 birds/group and fed insect-protected Bt or non-Bt corn for 42 days (Piva et al. 2001b). The insect-protected Bt and non-Bt hybrids were grown in three separate geographical locations and location was included as a test variable. The diets were composed of approximately 50% corn. There were no differences in feed intake or feed efficiency between the groups. At two of the locations, birds from the insect-protected *Bt* groups had slightly, but statistically higher live weight than the birds fed non-*Bt* corn. The weight difference was attributed to a 72% reduction in fumonisin levels in the insect-protected *Bt* corn compared with the non-*Bt* corn.

8.10.1.2 Lactating Cows

Green chop from insect-protected *Bt* or non-*Bt* corn was fed to groups of 12 lactating Holstein cows for 14 days. Cows on the three diets produced the same amount of milk (approximately 38 kg/day) throughout the study. In addition, no *Bt* protein was detected in milk from cows receiving the *Bt* diets (Faust and Miller 1997).

Mayer and Rutzmoser (1999) fed rations containing *Bt* or non-*Bt* corn silage (18 kg/day) to Fleckvieh breed cattle in Germany. Two groups of 12 cows were allocated to 5-week feeding periods in a crossover design. Silage feeding values were not different between the hybrids. In addition, silage source did not affect cattle performance, as measured by intake, milk production and milk composition (fat, protein, lactose, urea, somatic cell count).

Folmer et al. (2000a) compared the ruminal fermentation parameters and lactational performance between four balanced diets containing insect-protected *Bt* or non-*Bt* corn silage from either early- or late-maturity hybrids in a 4×4 Latin square design Diets contained 40% silage and 28% grain from the same corn hybrid plus 10% alfalfa silage, and 22% protein, mineral and vitamin supplement. No differences were detected between insect-protected *Bt* and non-*Bt* diets at either maturity for rumen fermentation characteristics (in-situ NDF (neutral detergent fiber) digestibility, rumen VFA (volatile fatty acids) concentration, rumen pH), milk production or milk composition. Animals fed early maturity hybrids (*Bt* and non-*Bt* combined) did have improved ($p < 0.005$) total rumen VFA and efficiency of production compared with later-maturity hybrids (*Bt* and non-*Bt* combined).

In one recent long-term study (Barriere et al. 2001), two groups of 24 lactating Holstein cows were fed silage from *Bt* or non-*Bt* hybrids for 13 weeks. Body weight, FCM (31.3 and 31.4 kg/day), milk protein (31.7 and 31.6 g/kg) and milk fat (36.7 and 37.0 g/kg) were not affected by corn hybrid. Cattle fed *Bt* silage had higher intakes than controls. Barriere et al. (2001) also reported on cheese-making qualities of milk from mid-lactation Holsteins fed *Bt* or non-*Bt* corn silage for 2- to 3-week intervals. There were no effects of corn source on protein fractions, fatty acid composition or coagulation properties of milk.

8.10.1.3 Beef and Sheep

Gestating beef cows were allotted to crop residue fields from which *Bt* (three different hybrids) or non-*Bt* corn was harvested (Russell et al. 2000). Groups of three cows were allowed to strip-graze paddocks within the fields for 126 days. Six similar cows were allotted to replicate dry lots. Corn borer infestations were greater in the non-*Bt* hybrid than in the three insect-protected *Bt* lines, although grain yield and dropped ears did not differ among hybrids. No differences were detected between residues from insect-protected *Bt* hybrids and non-*Bt* corn for dry matter or organic matter or in vitro digestible dry matter (IVOMD) yields at initiation of grazing. Dry matter, organic matter and IVOMD losses were greater from two of the insect-protected *Bt* hybrids than from the other two hybrids. Over the grazing season, no differences were detected between hybrids for composition rates of change. Amounts of hay required to maintain body condition score by grazing cows were 836 kg dry matter/cow less than those maintained in dry lots, but did not differ between hybrids.

To examine grazing preference and daily gain, 67 beef steers were allowed to graze *Bt* and non-*Bt* residue for 70 days while 16 additional steers were allowed to select grazing on *Bt* or non-*Bt* fields (Folmer et al. 2000b). Daily gain of steers was not different between the hybrid residues, averaging 0.28 kg/day. Also, steers given a choice had no grazing preference. European corn-borer pressure was minimal during the growing season and subsequently yields and residue were similar between the fields.

In a 2-year study, dry, pregnant beef cows (40 in year 1 and 36 in year 2) were allotted to graze either *Bt* or non-*Bt* corn residues for 34–42 days (Hendrix et al. 2000). During the first year, 20 additional cows were given access to either *Bt* or non-*Bt* fields, and observed three times daily for field preference. Average body weight change did not differ between cows grazing *Bt* or non-*Bt* fields. Cows given a choice of fields tended to graze as a group in variable patterns. Over the entire observation period 46% of cows were observed on the *Bt* field compared with 54% on the non-*Bt* field.

Daenicke et al. (1999) studied digestibility and animal performance in sheep and growing Holstein bull calves fed *Bt* or non-*Bt* corn silage. There were no differences in nutrient composition between the two silages. Using four sheep per group, there were no differences in digestibility of organic matter, fat, fiber or nitrogen-free extract between *Bt* and non-*Bt* silages. Two groups of 20 male Holstein calves (starting body weight of 188 kg at 165 days of age) were fed *Bt* or non-*Bt* corn silage until they attained a body weight of 550 kg. There were no differences in intake, body weight gain, feed conversion, hot carcass weight, dressing percentage or abdominal fat between calves fed *Bt* or non-*Bt* corn silage. Total dry matter intake and metabolizable energy (ME) were slightly lower in calves given *Bt* silage than in controls.

8.10.1.4 Swine

Grower-finisher performance and carcass characteristics from pigs fed insect-protected *Bt*, the non-*Bt* isogenic counterpart or commodity-sourced (CS) corn were compared. No differences were detected in average daily gain, average feed intake, or feed efficiency between pigs fed any of the three corn sources. Pigs fed insect-protected *Bt* or the non-*Bt* corn were not different in carcass weight, however pigs fed CS corn had heavier carcass weights and higher ($p < 0.05$) dressing percentages than the other two groups. Pigs fed the isogenic control had lower ($p < 0.05$) percent lean, greater back-fat depth at the tenth rib and last-rib location than pigs fed diets containing insect-protected *Bt* or the CS corn. Furthermore, pigs fed the non-*Bt* corn had greater ($p < 0.05$) back-fat depth at the last lumbar vertebrae than pigs fed CS corn. Marbling scores were highest ($p < 0.05$) in pigs fed insect-protected *Bt* and near isogenic control corn. The authors concluded that insect-protected *Bt* corn had no adverse effects on growth performance or carcass characteristics of grower-finisher pigs (Weber et al. 2000).

Weaned piglets were assigned to groups of 32 (16 castrated males and 16 females) and fed insect-protected *Bt* or non-*Bt* corn for 35 days (Piva et al. 2001a). The insect-protected *Bt* and non-*Bt* hybrids were grown in three separate geographical locations and each location was tested in this study. There were no differences between groups in feed intake or feed efficiency. Final live-weight and weight gain were slightly, but statistically significantly higher for the insect-protected *Bt* group than the non-*Bt* group. Female pigs fed insect-protected *Bt* corn had increased feed intake during days 15–35 of the study compared to controls. Intake and body weight differences may be due to the lower (69%) fumonisin levels in insect-protected *Bt* corn than in non-*Bt* corn.

Two digestible-energy experiments compared the nutritional value of insect-protected *Bt* and non-*Bt* corn as well as three non-transgenic commercial corn hybrids (Gaines et al. 2001b). Twenty cross-bred barrows were assigned randomly to one of five groups; fecal material and feed were analyzed to determine digestible energy coefficients over a 5-day feeding period. The experiment was repeated with 20 additional barrows. There were no differences in digestible energy between the insect-protected *Bt* and non-*Bt* near isogenic control, although differences were detected between the three commercial non-transgenic hybrids.

8.11 Cottonseed

The effect of feeding lactating dairy cows with cottonseed containing combined traits for insect protection (Cry1Ac protein) and tolerance to Roundup herbicide were compared in cattle fed two commercial non-transgenic cottonseed varieties as well as the parental line from which the transgenic line was derived (Castillo et al. 2001b). Twelve lactating, multiparous, Argentinean Holsteins, weighing on average 570 kg, were used in a ×4 Latin square design with three squares each containing four cows; there were four 4-week periods and four treatments. The cows consumed approximately 2.3 kg (dry matter basis) of cottonseed per day. There were no significant differences ($p > 0.05$) in dry matter intake, cottonseed intake, milk yield or composition (fat, protein, lactose, non-fat solids, urea) or body condition score.

The effect of feeding lactating dairy cows with cottonseed containing combined traits for insect protection (Cry1Ac and Cry2Ab proteins) (BollGard II) were compared in cattle fed cottonseed containing the Cry 1Ac protein alone (BollGard I), cottonseed containing the Roundup Ready (CP4 EPSPS) trait, or the non-transgenic control line from which the insect protected lines were derived (Castillo et al. 2001a). Twelve lactating, multiparous, Argentinean Holsteins weighing on average 570 kg were used in a 4×4 Latin square design as described previously. The cows consumed approximately 2.3 kg (dry matter basis) of cottonseed per day. There were no significant differences ($p > 0.05$) in dry matter intake, cottonseed intakes, milk yield or composition (fat, protein, lactose, non-fat solids, urea) or body condition score.

8.12 Conclusions

The introduction of Cry proteins into plants to provide protection against corn borers or cotton bollworms is a major innovation for agriculture, providing growers with an alternative to conventional chemical insecticides. The incorporated Cry proteins are effective in controlling target-insect pests since protection can be provided throughout the entire growing season. The Cry insect-control proteins have an unparalleled safety record following 40 years of use in agriculture. Their remarkable safety profile is a consequence of their highly specific mode of action and absence of toxicity to non-target organisms. The environment can benefit by using Cry proteins, which provide selective control of targeted pests and spare non-target organisms, thereby significantly reducing the use of some chemical insecticides. Because of reduced physical damage to the plant, insect protected corn also has been shown to have a reduced potential for contamination with fungi that produce fumonisins. These mycotoxins may negatively impact human or animal health when levels in corn exceed action levels that have been established by mycotoxin experts throughout the world. Although there is a compelling body of information to support the safety of insect protected plants for food/feed consumption, additional farm-animal feeding studies have been undertaken to compare the feeding value of these new varieties to existing commercial crops. These studies have been conducted in a variety of farm animal species. Swine and especially poultry are sensitive species to test for the nutritional adequacy of corn varieties; the absence of any adverse effects in the performance of swine and poultry fed insect-protected corn attests to the safety and nutritional value of these products. In head-to-head comparisons, the insect- protected plants supported animal growth and performance similar to their conventional counterparts. Thus, available data demonstrate the benefits and safety of insect-protected crops for human and animal consumption.

References

Aulrich K, Halle I, Flachowsky G (1998) Inhaltsstoffe und Verdaulichkeit von Maiskπrnen der Sorte Cesar und der gentechnisch veränderten Bt-hybride bei Legenhennen. Proc Einfluss von Erzeugung und Verarbeitung auf die Qualität laudwirtschaftlicher Produkte:465–468.

Barriere Y, Verite R, Brunschwig P, Surault F, Emile JC (2001) Feeding value of corn silage estimated with sheep and dairy cows is not altered by genetic incorporation of Bt176 resistance to *Ostrinia nubilalis*. J Dairy Sci 84:1863–1871

Baum, JA, Johnson TB, Carlton BC (1999) *Bacillus thuringiensis* natural and recombinant bioinsecticide products. In: Hall FR, Mean JJ (eds) Methods in biotechnology, vol 5. Biopesticides: use and delivery. Humana Press, Inc, Totowa, NJ, pp 189–209

Berberich SA, Ream JE, Jackson TL, Wood R, Stipanovic R, Harvey P, Patzer S, Fuchs RL (1996) Safety assessment of insect-protected cotton: the composition of the cottonseed is equivalent to conventional cottonseed. J Agric Food Chem 41:365–371

Betz FS, Hammond BG, Fuchs RL (2000) Safety and advantages of *Bacillus thuringiensis*-protected plants to control insect pests. Regulatory Toxicol Pharmacol 32:156–173

Brake J, Vlachos D (1998) Evaluation of event 176 "Bt" corn in broiler chickens. J Poultry Sci 77: 648–653

Cahagnier B, Melcion D (2000) Mycotoxines de *Fusarium* dans le mais-grains a la recolte; relation entre la presence d'insectes (pyrale, sesamie) et la teneur en mycotoxines. In: Piva G, Masoero F (eds) Proceedings of the 6th International Feed Conference, Food safety: current situation and perspectives in the European community. Piacenza, Italy 27–28 November, 2000, pp 237–249

Castillo AR, Gallardo MR, Maciel M, Giordano JM, Conti GA, Gaggiotti MC, Quaino O, Gianni C, Hartnell GF (2001a) Effect of feeding dairy cows with either BollGard ®, BollGard ® II, Roundup Ready ® or control cottonseeds on feed intake, milk yield and milk composition. J Dairy Sci 84(Suppl 1):Abstract 1712

Castillo AR, Gallardo MR, Maciel M, Giordano JM, Conti GA, Gaggiotti MC, Quaino O, Gianni C, Hartnell GF (2001b) Effect of feeding dairy cows with cottonseeds containing BollGard and Roundup Ready genes or control non-transgenic cottonseeds on feed intake, milk yield and milk composition. J Dairy Sci 84 (Suppl 1):Abstr 1713

Clark JH, Ipharraguerre IR (2001) Livestock performance: feeding biotech crops. J Dairy Sci 84 (E Suppl):E9-E18

Crickmore N, Ziegler DR, Feitelson J, Schnepf E, Van Rie J, Lereclue R, Baum J, Dean DH (1998) Revision of the nomenclature for the *Bacillus thuringiensis* pesticidal crystal proteins. Microbiol Mol Biol Rev 62:807–813

Culpepper AS, York AC (1998) Weed management in glyphosate-tolerant cotton. J Cotton Sci 4:174–185

Daenicke R, Gadeken D, Aulrich K (1999) Einsatz von Silomais herkömmlicher Sorten und der gentechnisch veränderten Bt Hybriden in der Rinderfütterung – Mastrinder. 12. Maiskolloquium:40–42

Dowd P (2000) Indirect reduction of era molds and associated mycotoxins in *Bacillus thuringiensis* corn under controlled and open field conditions: utility and limitations. J Econ Entomol 93:1669–1679

English L, Slatin SL (1992) Mode of action of delta-endotoxin from *Bacillus thuringiensis*: a comparison with other bacterial toxins. Insect Biochem Mol Biol 22:1–7

EPA (1995a) Fact Sheet for *Bacillus thuringiensis* subspecies *kurstaki* Cry1 A(b) delta endotoxin and its controlling sequences in corn. March 21 (Ciba Seeds)

EPA (1995b) Fact Sheet for *Bacillus thuringiensis* subspecies *tenebrionis* Cry3 A delta endotoxin and its controlling sequences in potato. May 5 (Monsanto)

EPA (1996) Fact Sheet for *Bacillus thuringiensis* subspecies *kurstaki* Cry1A(b) delta endotoxin and its controlling sequences as expressed in corn. December 20 (Monsanto)

EPA (1998a) EPA Registration Eligibility Decision (RED) *Bacillus thuringiensis*. EPA 738-R-98-004, March (1998)

EPA (1998b) (RED Facts) *Bacillus thuringiensis*. EPA-738-F-98-001

FAO/WHO (2001) Joint FAO/WHO expert committee on food on food additives. Fifty-sixth meeting. Geneva, 6–15 February. http://www.who.int/pcs/jecfa/jecfa.htm

Faust M, Miller L (1997) Study finds no Bt in milk. IC-478. Fall special livestock edition. Iowa State University Extension, Ames, Iowa, pp 6–7

Fischhoff DA, Bowdish KS, Perlak FJ, Marrone PG, McCormick SM, Nidermeyer JG, Dean DA, Kusano-Kretzmer K, Mayer EJ, Rochester DE, Rogers SG, Fraley RT (1987) Insect tolerant transgenic tomato plants. Bio/Technology 5:807–813

Folmer JD, Grant RJ, Milton CT, Beck JF (2000a) Effect of Bt corn silage on short-term lactational performance and ruminal fermentation in dairy cows. J Dairy Sci 83 (5):1182 Abstract 272

Folmer JD, Erickson GE, Milton CT, Klopfenstein TJ, Beck JF (2000b) Utilization of Bt corn residue and corn silage for growing beef steers. J Animal Sci 78 (Suppl 2):67

Gaines AM, Allee GL, Ratliff BW (2001a) Nutritional evaluation of Bt (MON810) and Roundup Ready corn compared with commercial hybrids in broilers. Poultry Sci 80 (Suppl 1):Abstr 214

Gaines AM, Allee GL, Ratliff BW (2001b) Swine digestible energy evaluations of Bt (MON810) and Roundup Ready corn compared with commercial varieties. J Anim Sci 79 (Suppl 1):109 Abstr 453

Gianessi LP, Carpenter JE (1999) Agricultural biotechnology: insect control benefits. National Center for Food and Agricultural Policy, Washington DC

Halle I, Aulrich K, Flachowsky G (1998) Einsatz von Maiskörnern der Sorte Cesar und des gentechnisch veränderten Bt-Hybriden in der Broiler mast. Proc 5. Tagung, Schweine- und Geflügelernährung, 01.-03.12.1998, Wittenberg, pp 265–267

Hendrix KS, Petty AT, Lofgren DL (2000) Feeding value of whole plant silage and crop residues from Bt or normal corns. J Anim Sci 78(Suppl 1):273 Abstract

Hofmann C, Luthy P, Hutter R, Pliska V (1988) Binding of the delta endotoxin from *Bacillus thuringiensis* to brush-border membrane vesicles of the cabbage butterfly (*Pieris brassicae*). Eur J Biochem 173:85–91

Hofte H, Whitely HR (1989) Insecticidal crystal proteins of *Bacillus thuringiensis*. Microbiol Rev 53:242–255

IPCS (2000) International programme on chemical safety – environmental health criteria 217: *Bacillus thuringiensis*. http://www.who.int/pcs/docs/ehc_217.html

James C (1997) Global status of transgenic crops in (1997) ISAAA Briefs No. 5. ISAAA, Ithaca, NY

James C (1998) Global review of commercialized transgenic crops: 1998. ISAAA Briefs No. 8. ISAAA, Ithaca, NY

James C (1999) Preview: global review of commercialized transgenic crops: 1999. ISAAA Briefs No 12. ISAAA, Ithaca, NY

Klotz-Ingram C, Jans S, Fernandez-Cornejo J, McBride W (1999) Farm-level production effects related to the adoption of genetically modified cotton for pest management. AgBioForum 2 (2):73–84; retrieved July 15, 1999 from the world wide web: http://www.agbioforum.missouri.edu

Knowles BH, Ellar DJ (1987) Colloid-osmotic lysis is a general feature of the mechanisms of action of *Bacillus thuringiensis* (delta)-endotoxins with different insect specificity. Biochim Biophys Acta 924:509–518

Marasas WFO, Jaskiewicz K, Venter FS, van Schalkwyk DJ (1988) Fusarium moniliforme contamination of maize in oesophageal cancer areas in the Transkei. S Afr Med J 74:110–114

Marra M, Carlson G, Hubbell B (1998) Economic impacts of the first crop biotechnologies. Available on the world wide web: http://www.ag.econ.ncsu.edu/faculty/marra/online.html

Mayer J, Rutzmoser K (1999) Einsatz von silomais herkömmlicher sorten und der gentechnisch veranderten Bt-hybriden in der Rinderfutterung: bei Milchkuhen. 12 Maiskolloquium:36–39

McClintock JT, Schaffer CR, Sjoblad RD (1995) A comparative review of the mammalian toxicity of *Bacillus thuringiensis*-based pesticides. Pestic Sci 45:95–105

Mireles A Jr, Kim S, Thompson R, Amundsen B (2000) GMO (Bt) corn is similar in composition and nutrient availability to broilers as non-GMO corn. J Poultry Sci 79 (Suppl 1):65–66 Abstr 285

Munkvold GP, Hellmich RL, Rice LG (1999) Comparison of fumonisin concentrations in kernels of transgenic Bt maize hybrids and nontransgenic hybrids. Plant Dis 83:130–138

Norred WP (1993) Fumonisins – mycotoxins produced. J Toxicol Environ Health 38:309–328

Noteborn HPJM, Rienenmann-Ploum ME, van den Berg JHJ, Alink GM, Zolla L, Kuiper HA (1993) Food safety of transgenic tomatoes expressing the insecticidal crystal protein Cry1Ab from *Bacillus thuringiensis* and the marker enzyme APH(3′)II. Med Fac Landbouww Univ Gent 58/4b

Noteborn HPJM, Rienenmann-Ploum ME, van den Berg JHJ, Alink GM, Zolla L, Kuiper HA (1994) Consuming transgenic food crops: the toxicological and safety aspects of tomato expressing Cry1Ab and NPTII. ECB6. Proceeding of the 6th European Congress on biotechnology. Elsevier, Amsterdam

Perlak FJ, Deaton RW, Armstrong TA, Fuchs RL, Sims SR, Greenplate JT, Fischhoff DA (1990) Insect resistant cotton plants. Bio/Technology 8:939–943

Pietri A, Piva G (2000) Occurrence and control of mycotoxins in maize grown in Italy. In: Piva G, Masoero F (eds) Proceedings of the 6th International Feed Conference, Food safety: current situation and perspectives in the European community. Piacenza, Italy, 27–28 November, pp 226–236

Piva G, Morlacchini M, Pietri A, Piva A, Casadei G (2001a) Performance of weaned piglets fed insect-protected (MON810) or near isogenic control corn. J Anim Sci 79 (Suppl 1):106 Abstr 1324

Piva G, Morlacchini M, Pietri A, Rossi F, Prandini A (2001b) Growth performance of broilers fed insect-protected (MON810) or near isogenic control corn. Poultry Sci 80 (Suppl 1):320 Abstr 1324

Russell JR, Hersom MJ, Pugh A, Barrett K, Farnham D (2000) Effects of grazing crop residues from Bt-corn hybrids on the performance of gestating beef cows. Abstract 244 presented at the Midwestern Section ASAS and Midwest Branch ADSA 2000 Meeting, Des Moines, IA. J Anim Sci 78 (Suppl 2):79–80

Sanders PR, Lee TC, Groth ME, Astwood JD, Fuchs RL (1998) Safety assessment of the insect-protected corn. In: Thomas JA (ed) Biotechnology and safety assessment, 2nd edn. Taylor and Francis, London, pp 241–256

Sjoblad RD, McClintock JT, Engler R (1992) Toxicological considerations for protein components of biological pesticide products. J Econ Entomol 80:717–723

Spencer TM, Orozco EM, Doyle RM (1996) October 14, 1986. Petition for determination of non-regulated status: insect protected corn (*Zea mays* L.) with *cry1Ac* gene from *Bacillus thuringiensis* subsp. *kurstaki*. DEKALB Genetics Corporation, Mystic, Connecticut

Taylor ML, Harnell GF, Nemeth MA, George B, Astwood JD (2001) Comparison of broiler performance when fed diets containing YieldGard corn, YieldGard and Roundup Ready corn, parental lines or commercial corn. Poultry Sci 80 (Suppl 1):319 Abstr 1321

Vaeck M, Reybnaerts A, Hofte J, Jansens S, DeBeuckeleer M, Dean C, Zabeau M, Van Montagu M, Leemans J (1987) Transgenic plants protected from insect attack. Nature 328:33–37

Weber TE, Richert BT, Kendall DC, Bowers KA, Herr CT (2000) Grower-finisher performance and carcass characteristics of pigs fed genetically modified "Bt" corn. Purdue University 2000 Swine Day Report, http://www.ansc.purdue.edu/swine/swineday/sday00/psd07–2000.html

9 Safety Assessment of Genetically Modified Rice and Potatoes with Soybean Glycinin

K. Momma, W. Hashimoto, S. Utsumi, and K. Murata

9.1 Introduction

Following the rational development of genetic engineering, many genetically modified (GM) crops are now being exploited in the world. However, the safety of GM crops regarding public health and the environment has not been thoroughly ensured, and recent reports described the possibility of homologous recombination between viral RNA and transgenic plant transcripts (Green and Allison 1994), gene flow from GM crops to weeds (Jenczewski et al. 1999), high allergenicity of GM *Canola* and GM *Vicia* (bean) (Nordlee et al. 1996), growth retardation and suppression of the rat immune system by GM potatoes (Enserink 1999), harmful effect of pollen from pest-resistant GM corn on non-target pests (Losey et al. 1999), and allergenicity of pest-resistant corn (Star-Link) (http://www.mhlw.go.jp/search/mhlwj/mhlw/houdou/). These examples have apparently raised issues related to food safety, regulatory oversight and consumer acceptance of biotechnology-derived foods, although some, i.e. not all, of the results were obtained under insufficient experimental conditions and/or were overestimated, as reported by Ewen and Pusztai (1999) in the case of the effect of GM potatoes on the rat immune system.

The safety of GM crops and GM foods derived from them has been viewed from the standpoint of either the *process* or the *product*. The *process* is thought to involve neither essentially nor potentially dangerous steps, since gene engineering technology is considered to be more precise and intentional than traditional genetic techniques such as hybridization and mutagenesis. It has therefore been concluded that safety studies should be limited to the *product*, and the safety of GM crops has been assessed based on the scientific concept of *substantial equivalence*, which was coined by the Organization for Economic Cooperation and Development (OECD) in 1993, and recommended by the US Food and Drug Administration (FDA) policy. This concept is based on the demonstration of substantial equivalence between a GM crop and its natural antecedent. The safety defined by the concept of substantial equivalence represents nutritional, biochemical and morphological equivalence between GM crops and their counterparts, and several kinds of GM crops have been proved to be safe based on this concept, and some of them have been commercialized or nearly so.

Commercialized GM crops such as herbicide-resistant soybeans and pest-resistant corn have been developed by introducing bacterial genes. These crops

Molecular Methods of Plant Analysis, Vol. 22
Testing for Genetic Manipulation in Plants
Edited by J.F. Jackson, H.F. Linskens, and R.B. Inman
© Springer-Verlag Berlin Heidelberg 2002

were usually produced to increase the productivity, and in many cases the expression levels of proteins specified by the bacterial genes are low, being approximately 0.01 to ~0.1% of the total proteins in the host crops. We designate these GM crops or GM foods, which promised benefits for farmers, as *first generation GM crops* (*or GM foods*). However, it has been revealed that biotechnology impacts the entire food chain, from the genetic improvement of agricultural crops to the processing, packaging and distribution of processed foods. In particular, there is increasing evidence that GM crops can produce healthier foods, and be used as chemical factories for the production of construction materials for the medical, food, chemical and agricultural industries.

Therefore, the next major step of plant engineering is the creation of crops with apparent and direct benefits for consumers. Namely, the trend in GM crop creation is apparently changing from the improvement of productivity to that of quality, and the creation of crops with high nutritional value or having the ability to prevent life-style related diseases (diabetes, hypertension, hypercholesterolemie, allergy, corpulence, etc.) and infection by pathogenic bacteria (plant vaccines) has been considered. These physiologically, nutritionally and/or medically functional novel crops, which we designate as *second generation GM crops* (*or GM foods*), to distinguish them from the first generation GM crops (or GM foods) defined above, are often created by means of artificially designed genes and/or multi-gene systems. In these crops, the expression levels of proteins specified by the genes or gene-systems are, under certain circumstances, significantly high, reaching 5–10% (Katsube et al. 1999). The high level expression of foreign proteins makes GM crops basically different from their natural counterparts, and the safety assessment standard hitherto used, substantial equivalence, is not necessarily adequate for safety assessment of second generation GM crops (or GM foods).

Thus, biotechnology is rapidly expanding to include even more sophisticated, precise and rapid methods for the construction and analysis of GM crops and foods that reach well beyond current capabilities for modifying the food supply. Industry and agriculture, as well as human society, now face the problem of how to regulate GM crops produced by means of biotechnology to improve the supply of sound, wholesome, tasty, convenient and affordable foods. Based on the results as to safety assessment of GM crops (rice and potatoes) with soybean glycinin, we discuss here potential problems hindering the public acceptance of GM crops and GM foods.

9.2 Safety Assessment of Genetically Modified Crops

Rice is one of the most important crops in the world, and intense efforts, including the use of gene engineering technology, have been made to increase its yield, nutritional value and/or physiological properties (Nakamura and Matsuda 1994; Itoh 1996; Gura 1999). However, the lysine content is low in ordinary rice and insufficient for children, particularly those under 2 years (FAO/WHO/UNU 1985). The potato tuber is currently considered to be an important food for the ingestion of selected vitamins and minerals such as vitamin B6, ascorbic acid and potassium (Lavirk et al. 1995). However, the levels of leucine, lysine and threonine in the storage proteins of potato tubers are low and insufficient for the nutritional needs of children under 5 years (FAO/WHO/UNU 1985). Potatoes are now genetically modified to produce highly qualitative proteins, pest-resistant proteins, or antigens for use as edible vaccines against diarrhea caused by *Escherichia coli*.

Glycinin is one of the major storage proteins of soybean and consists of six subunits, each of which is composed of acidic and basic polypeptides. This protein shows physiological [e.g. lowering of the serum cholesterol level (Sugano et al. 1990; Kito et al. 1993)] and physicochemical [e.g. emulsification and heat-inducible gelation (Kim et al. 1990)] properties. In order to boost the low amino acid contents of rice and potatoes, and to improve their physiological and physicochemical properties simultaneously, soybean glycinin genes have been introduced into rice (Katsube et al. 1999) and potatoes (Utsumi et al. 1994). The GM rice and GM potatoes thus created are expected to be sound and wholesome, and may serve as second generation GM crops that are particularly effective for individuals suffering from hypercholesterolemia.

The quality and safety of GM rice and GM potatoes with soybean glycinin have been assessed by means of component analyses, and short- and long-term animal feeding experiments. The analyses were carried out with regard to more than sixty components belonging to nutritionally and physiologically important classes of compounds: macronutrients [carbohydrates, lipids (12 fatty acids) and proteins (18 amino acids)], micronutrients (5 minerals and 5 vitamins), and others (fiber and ash).

9.2.1 Genetically Modified Rice

cDNA of the soybean glycinin gene was placed between the rice glutelin promoter and terminator (Fig. 9.1A), and the resultant DNA was introduced into rice (strain: *Matsuyamamii*) by electroporation. The transformed rice was isolated utilizing a biaphorus gene marker (Katsube et al. 1999). Seeds of the control (non-transgenic) and GM rice were planted in pots placed in a closed greenhouse and then cultivated for 6 months. There were no significant differences in shape or weight between the control and GM rice grains (Momma et al. 1999).

Expression and Digestibility. The expression level of soybean glycinin in the total proteins of GM rice was 4 to ~5%, i.e. 40 to ~50 mg/g-protein (Momma et al. 1999; Fig. 9.2). Glycinin in the GM rice was promptly and completely digested by simulated gastric and intestinal fluid (Momma et al. 1999; Fig. 9.2).

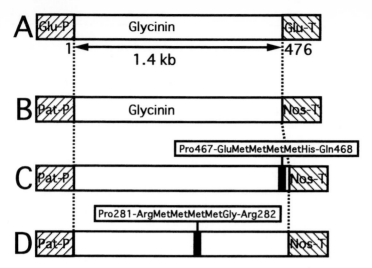

Fig. 9.1A–D. Structures of native and modified soybean glycinins in GM rice and GM potatoes (Hashimoto et al. 1999a; Momma et al. 1999). The cDNA (1.4 kb) coding for native glycinin (476 amino acids) inserted between the promoter (Glu-P) and terminator (Glu-T) for the rice glutelin gene was electroporated into rice (**A**). Native (**B**) and modified (**C, D**) glycinin cDNAs placed between the promoter for the patatin gene (Pat-P) and the terminator for the nopaline synthase gene (Nos-T) were transferred into potatoes by the use of *Agrobacterium tumefaciens*. In Ag891 (**C**) and Ag921 (**D**), tetramethionyl residues were introduced between Pro-467 and Gln-468, and Pro-281 and Arg-282 of glycinin, respectively

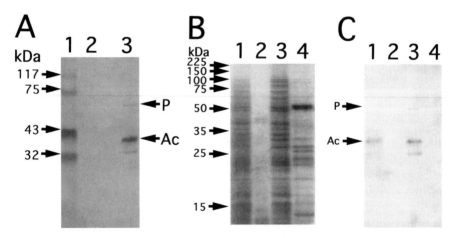

Fig. 9.2A–C. Electrophoretic profiles of soybean glycinin expressed in rice (Momma et al. 1999). **A** Expression of glycinin in GM rice. Proteins extracted from GM and non-GM rice were subjected to SDS-PAGE, followed by immunostaining with anti-glycinin antibodies. *Lane 1* Molecular weight standards, *lane 2* non-GM rice, *lane 3* GM rice. **B, C** Digestibility of glycinin expressed in GM rice. Proteins extracted from GM rice were incubated with simulated gastric (SGF) and intestinal (SIF) fluids. At the prescribed times, a sample was taken and subjected to SDS-PAGE, followed by protein staining (**B**) and immunostaining with anti-glycinin antibodies (**C**). *Lane 1* 0 min (SGF), *lane 2* 10 min (SGF), *lane 3* 0 min (SIF), *lane 4* 30 min (SIF). *P* and *Ac* indicate the precursor and acidic subunit of glycinin, respectively

Protein Fluctuations. The GM rice was different from the control rice in the amounts and kinds of protein species. The total protein in the GM rice (8.0/100g) was about 20% higher than that in the control rice (6.5/100g; Momma et al. 1999). The increased level of protein observed in the GM rice was presumably due to the marked fluctuation of protein species in addition to the glycinin expressed. One protein (32 kDa) present in the control rice was absent in the GM rice.

Metabolite Fluctuations. Accompanying the higher protein level in the GM rice, the contents of almost all amino acids including lysine were higher (20% more) in the GM rice (Momma et al. 1999). The levels of fatty acids remained unchanged, except for increases in palmitic acid (15% more), linoleic acid (12% more), and 18:1 fatty acid (11% more) in the GM rice (Momma et al. 1999). No significant differences were found in the contents of minerals and vitamins between the control and GM rice, except for that of vitamin B6, which was appreciably higher (50% more) in the GM rice (Momma et al. 1999). No significant differences were observed in the contents of fructose, glucose, sucrose and maltose between the control and GM rice (Fukuda et al., unpubl.).

Allergenicity. The rice used as the host for the glycinin gene contains four inherent allergenic proteins [I (14–16 kDa), II (26 kDa), III (33 kDa) and IV (56 kDa)] (Urisu et al. 1991). Protein III was recently identified as an enzyme, glyoxalase I, catalyzing the conversion of toxic compound methylglyoxal to *S*-lactoylglutathione in the presence of glutathione (Matsuda, unpubl.). The amount of each allergenic protein (I, II, III and IV) was not affected by the introduction of soybean glycinin genes, their levels being comparable with those in the control rice (Matsuda, unpubl.).

In Vivo Safety. To check the effects of the metabolic fluctuations described above, GM rice was subjected to animal feeding experiments for 4 weeks (Momma et al. 2000). The administered amount was 10 g/kg-rat/day, which is the maximum volume that can be given and ten times higher than that prescribed for the safety assessment of food additives. No abnormalities were observed in the red blood cell, leukocyte and hemoglobin counts, or in the hematocrit, platelets, mean cell volume, mean cell hemoglobin, mean cell hemoglobin concentration, prothrombin time, activated partial thromboplastin time, fibrinogen, and reticulocyte properties; the activities of aspartic acid aminotransferase (GOT), alanine aminotransferase (GPT), alkaline phosphatase (ALP), and γ-glutamyltranspeptidase (γ-GTP); the contents of calcium, sodium, potassium, phosphorus and chloride; or the amounts of total cholesterol, total protein, triglyceride, creatine, total bilirubin, urea nitrogen, inorganic phosphate and glucose. No abnormalities of organs (brain, pituitary, salivary glands, thyroids, thymus, lungs, heart, liver, spleen, kidneys, adrenals, testes and prostate) were observed regarding shape and function. No biochemical, nutritional or morphological abnormalities were detected in long-term chronic toxicity experiments (unpublished data). The ability of GM rice to induce deformity, mutagenicity and cancer is now being examined.

9.2.2 Genetically Modified Potatoes

Soybean glycinin contains variable regions in the middle and carboxy-terminal of its molecule (Kim et al. 1990; Utsumi et al. 1994). To increase the sulfur amino acid content of potatoes, the nucleotide sequence for tetra-methionyl residues was inserted into the C-terminal and middle variable regions of the coding sequence of the intact glycinin gene, and the resultant genes were placed between the potato patatin gene-promoter and nopaline synthase gene terminator sequences of a vector plasmid carrying the neomycin phosphotransferase II gene (Fig. 9.1). The vector and other constructs containing glycinin were introduced into potatoes [strain: *May Queen* (MQ)] using *Agrobacterium tumefacience* as a gene shuttle, and the resultant transformants were designated as P127, Ag877, Ag891 and Ag921 (Fig. 9.1). MQ and P127 were used as controls (Kim et al. 1990; Utsumi et al. 1994).

There were no significant differences in appearance between the control (MQ and P127) and GM (Ag877, Ag891 and Ag921) potatoes (Hashimoto et al. 1999a). Although neomycin phosphotransferase II, a selection marker, was produced by the GM potatoes, including P127, the safety of the protein has been confirmed by the US Department of Agriculture, the FDA, and by the US Environmental Protection Agency.

Expression and Digestibility. The expression levels of soybean glycinin in GM potatoes (Ag877, Ag891 and Ag921) were estimated to be 1–3% (12, 31 and 23 mg/g-protein, respectively). The expressed product was present as proglycinin in the tubers of these GM potatoes (Fig. 9.3). Contrary to in the case of GM rice, no marked differences in the protein profile were observed between GM (Ag877, Ag891 and Ag921) and control (MQ and P127) potatoes, except for the glycinin expressed. The native and designed glycinins expressed in potatoes were completely digested by gastric and intestinal fluid (Hashimoto et al. 1999a).

Vector and Artificially Designed Genes. The vector plasmid and the designed glycinin genes, as well as the intact one, contain no CpG motifs. They all showed no effect on CHO K18 cell proliferation (Hashimoto et al. unpubl.).

Metabolite Fluctuations. Except for fatty acids, glycoalkaloids and sugars, no significant differences were observed in the contents of macro- or micronutrients between control (MQ and P127) and GM (Ag877, Ag891 and Ag921) potatoes (Hashimoto et al. 1999a). In GM potatoes including P127, a few fatty acids (12:0 and 20:1) absent in MQ were newly found. The contents of fatty acid (16:1) and unknown fatty acids increased, while fatty acid (17:0) was decreased

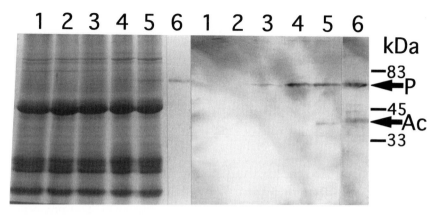

Fig. 9.3. Expression of soybean glycinin in GM potatoes (Hashimoto et al. 1999a) Proteins extracted from GM potato tubers were subjected to SDS-PAGE, followed by protein staining (*left*) and immunostaining with anti-glycinin antibodies (*right*). *Lane 1* MQ (control), *lane 2* P127, *lane 3* Ag877, *lane 4* Ag891, *lane 5* Ag921, *lane 6* glycinin. P and Ac indicate the precursor and acidic subunit of glycinin, respectively

Table 9.1. Levels of glycoalkaloids in potato tubers (mean±SEM) (from Hashimoto et al. 1999a)

Composition	Content (mg/kg)				
	MQ	P127	Ag877	Ag891	Ag921
α-Solanine	23.5 ± 0.5	54.0 ± 1.0	30.5 ± 0.5	37.0 ± 1.0	37.5 ± 1.5
α-Chaconine	53.0 ± 1.0	107 ± 1.0	69.0 ± 0.0	82.0 ± 3.0	74.5 ± 1.5

in all GM potatoes including P127 (Hashimoto et al. 1999a). However, the observed changes in the contents and components of fatty acids were too small to have any hazardous effects. Glycoalkaloids are toxic compounds present in potato tubers. Ninety-six percent of the glycoalkaloids comprise α-solanine and α-chaconine. The levels of both glycoalkaloids increased in all GM potatoes including P127 compared with in MQ (Hashimoto et al. 1999a) (Table 9.1). In particular, the contents of glycoalkaloids in P127 (160 mg/kg of potato) were twofold higher than those in MQ, thus suggesting the necessity of ensuring the safety of the vector itself. A similar increase in the glycoalkaloid level was observed in the case of pest-resistant GM potatoes (Lavirk et al. 1995). Since the safety of glycoalkaloids has been reported to be less than 200 mg/kg of potato (Smith et al. 1996), the increases in the glycoalkaloid levels observed in GM potatoes including P127 are considered to have no hazardous consequences. In MQ and P127 potatoes, the levels of sucrose, fructose, glucose and maltose were very similar (Fukuda et al., unpubl.). However, in one of the GM potatoes in the Ag891 series, the levels of fructose, glucose and sucrose were substantially (twofold) increased in comparison with those in controls (MQ and P127).

In Vivo Safety. GM potatoes were administrated to rats, and acute- and sub-acute toxicity was checked by comparison with rats similarly treated with a commercially obtainable diet (Hashimoto et al. 1999b). The administered amount chosen was 2 g/kg-rat/day, which is two times higher than that prescribed for the safety assessment of food additives. During the test period, no marked differences in body weight increase or food intake were observed between the control and GM potatoes. After administration for 4 weeks, the properties of blood and internal organs were examined. No abnormalities were found in the red blood cell, hemoglobin and leukocyte (lymphocytes, neutrophils, eosinophils, basophils, and monocytes) counts, or in other properties examined in the case of GM-rice assessment; the activities of GOT, GPT, and γ-GTP; the contents of calcium, sodium, potassium, phosphorus and chloride; or the amounts of total cholesterol, total protein, creatine, total bilirubin and glucose. The liver, kidneys and other organs also showed no pathological abnormalities in weight, shape or other properties. Although ALP activity in rats administered GM potatoes (Ag921) decreased to 80% of that in control rats similarly treated with MQ, these results as to acute and sub-acute levels of animal feeding experiments indicated that GM potatoes created by the use of native and designed soybean glycinin genes are as safe as non-GM potatoes (MQ).

9.3 Concluding Remarks

Genes that confer tolerance and/or resistance to herbicides or pests have been the target of genetic engineers for more than a decade. Following the successful molecular breeding of glyphosate-tolerant soybeans, attempts to make maize, soybeans, oilseed rape, cotton and sugar beet herbicide- and/or pest-resistant are underway in many countries. Some researchers are now turning their attention to so-called non-primary crops, and one such crop, the sweet potato, has already been genetically modified to improve its protein quality (Moffat 1998). Such GM prime and non-prime crops will be useful for enhancing the productivity of crops in the 21st century, in which we face an unpredictable population explosion.

The genetic traits currently being introduced into crops tend to be relatively simple and, in most cases, the crops are intact and monogenic. However, it is not easy to find genes suitable for improvement of the properties of crops. To overcome this limitation, attempts to use many genes (Moffat 1999), or artificially designed genes (Kim et al. 1990; Utsumi et al. 1994), or to express large amounts of foreign proteins in a single crop have been considered for the improvement of physiological functions. Indeed, an array of 7 to ~14 genes has already been put into one rice strain (Gura 1999; Moffat 1999). These methods are apparently fast and convenient ways to create target crops or foods. Recently developed DNA microarrays have made it possible to predict the functions of genes. This method facilitates the screening of genes responding to pathogens, pests, drought, cold, salt, herbicides and insecticides as well as physiological and food-biotechnological properties. By combining DNA microarrays with the gene design technique, crops and foods exhibiting higher productivity or a greater impact on physiological functions will be created.

Thus, the trend in GM crop construction is apparently towards increasing complexity and diversity, and the safety assessment of the resulting GM crops should improve with the advance of technology. Currently, the safety of GM crops regarding human health has been ensured based on the scientific concept of substantial equivalence. This concept requires that GM crops should be at least as safe as the traditional counterparts with a sufficient history of safe use. However, GM crops expressing large amounts of foreign proteins or physiologically and/or medically important proteins seem to be insubstantially equivalent to the traditional counterparts. In such crops, the cellular levels of toxic and/or physiologically important molecules are more or less influenced, as shown in the cases of GM rice and GM potatoes. These facts indicate that an introduced gene or the gene product may behave differently when working within its new host and therefore the original genetic properties of the host will be changed or disrupted. In particular, the introduction of many genes will result in more complex metabolic fluctuations than in the case of the introduction of a single gene. These fluctuations may bring about changes in nutritional value, allergenicity and other properties of the crops. Thus, for assessing GM crops with marked metabolic fluctuations, the concept of substantial equivalence does not hold good; thus more detailed safety assessment standards are needed which are based on the results of biological, toxicological and immunological tests analogous to the clinical trials used to assess the safety of drugs.

The safety assessment of first generation GM crops has been almost completely limited to acute- and sub-acute toxicological studies. However, it is anticipated that the safety assessment of second generation GM crops may require more detailed examination regarding their effects, especially on human health, although in addition to the problematic and expensive nature of toxicological and biochemical tests, there is some difficulty in extrapolation of the results of animal experiments to humans, since detoxification systems in mice and/or rats are largely different from those in humans in activity and amount, and in the detoxification enzyme species. It is also not easy to feed relevant doses of GM crops, although in our chronic toxicity examination we administered 4 g/kg-rat/day GM rice, which is four times higher than the dose prescribed for the safety assessment of food additives. To complement the results of animal feeding experiments, studies involving cultured cell systems may be effective for the in vivo safety assessment of GM crops, especially that of second generation GM crops (or GM foods).

As reasons for the metabolite fluctuations observed in the cases of GM rice and GM potatoes, the following two are considered. One is the effect of large amounts of foreign proteins expressed in host cells. The proteins may interfere with some metabolic pathways and induce a metabolic disturbance in the host cells. The other one is the so-called *position effect*. Although biotechnology is more precise and intentional than traditional breeding technology, we still can not control the sites of insertion of foreign genes in the nuclei of the host crops. This uncertainty entails unpredictable patterns of gene expression and genetic function. In fact, the specific and almost complete disappearance of a 14- to 16-kDa allergenic protein, possibly due to the position effect, has been observed in the case of GM rice with soybean ferritin (Matsuda, unpubl.). The development of a method for controlling and/or regulating the insertion sites in the nuclear DNA of host crops will undoubtedly facilitate the creation of GM crops with enhanced safety as foods.

These facts regarding fluctuations of metabolites and allergenic proteins indicate that the concept of substantial equivalence is a rather weak screen for various hazards caused by GM crops, especially by second generation GM crops (or GM foods), and that additional scientific data are required for confirmation of the safety of GM crops. Such data are as follows: (1) Data on the highly-sensitive two-dimensional electrophoretic profiles of proteins, (2) data on the *inherent allergenic proteins* in host crops and the changes in them after the introduction of foreign genes or artificially designed genes, (3) data on the allergenic properties of proteins specified by artificially designed genes, (4) data on the immunological and digestive properties of viral, bacterial and artificially designed genes, (5) data obtained on short- and long-term toxico-logical animal testing, (6) data on the chromatographic and spectrometric fingerprints of cellular metabolites in GM crops transformed with foreign genes or vector plasmids, and (7) data obtained in a prediction (scenario) study after the introduction of GM crops and/or GM foods as to human health, society and the environment.

In addition to the assessment of the safety of GM crops as to public health, the environmental safety must be monitored. One of the perplexing problems regarding environmental safety is gene pollution, especially that caused by gene flow from GM crops to other habitats. A recent report indicated the risks of pest-resistant potatoes or corn for non-target insects through pollen move-ment (Losey et al. 1999), although Losey et al. stated in their report "it would be inappropriate to draw any conclusions about the risk to Monarch popula-tions in the field based solely on these initial results". However, limited evi-dence available so far has undoubtedly left researchers apparently divided on the risk to non-target pests of crops transformed with genetic material from the bacterium *Bacillus thuringiensis* (*Bt*) (Losey et al. 1999). Some entomolo-gists found no differences in the numbers of beneficial insects when they sampled fields of *Bt* sweetcorn and non-*Bt* corn, but others demonstrated an effect in laboratory experiments. Thus, the results obtained so far regarding *Bt* toxin are contradictory (Shelton and Roush 1999; Wraight et al. 2000), and more systematic examination in the laboratory and in the field is indispensable for determining the effects of GM crops on biological systems, and for limiting or entirely preventing damage to non-target and beneficial insect species. Most crops are modified by inserting genes into the nucleus, and the genes can therefore be spread to other crops or wild relatives through pollen movement. One of the promising strategies for diminishing this risk is the use of chloroplast DNA (Daniell et al. 1998). As chloroplasts are presented in plant cells in multicopy numbers and are inherited maternally in many species, the use of chloroplasts is expected to prevent the problem of gene flow through pollen. In fact, an attempt to express *Bt* in chloroplasts has been made (Kota et al. 1999).

It is necessary that the risk assessment of GM crops and/or GM foods utilized as practical foods or food materials should be continued to ensure the safety of these novel crops. Along with risk assessment, the development of a prediction or scenario (bird's-eye view) study is required to predict the quality and magnitude of the effects of GM crops on public health and the environment. The results of the recent studies conducted by Watkinson et al. (2000) and Kubo and Saito (2000) may be significant. Watkinson et al. predicted a drastic decrease in the number of skylarks when GM herbicide-tolerant crops are cultivated. Kubo and Saito indicated the safety of nutrition-enriched GM crops (soybeans enriched with oleic acid, and rice enriched with β-carotene and iron) through a simulation based on the maximum intake of foods and the nutrition requirement threshold. Although these studies involved many hypothetical factors, such scenario studies will contribute to the safety assessment of GM crops.

Thus, the safety assessment of GM crops apparently requires an understanding of the molecular biology and biochemistry of plants, human physiology, public health, food chemistry and entomology, and the interdisciplinary collaboration of these sciences may be needed in order to ensure safety and a healthy food supply for the new century. Of course, we should be careful and modest in the creation of GM crops, since this act is able to more or less change nature. By assessing the safety of these novel crops regarding human health and the environment, we should create safe crops and promote the sound growth of DNA technology. Challenge in this direction regarding safety assessment will no doubt yield a wealth of information of a most unexpected nature on GM crops, providing a fruitful and satisfactory settlement of the problems regarding GM crops in the years to come.

Acknowledgements. This work was supported in part by a Grant-in-Aid from the Bio-oriented Technology Research Advancement Institution (BRAIN).

References

Daniell H, Datta R, Varma S, Gray S, Lee SB (1998) Containment of herbicide resistance through genetic engineering of the chloroplast genome. Nat Biotechnol 16:345–348
Enserink M (1999) Preliminary data touch off genetic food fight. Science 283:1094–1095

Ewen SW, Pusztai A (1999) Effect of diets containing genetically modified potatoes expressing *Galanthus nivalis* lectin on the rat small intestine. Lancet 354:1353–1354

FAO/WHO/UNU (1985) Factors affecting energy and protein requirements. In: Energy and protein requirements. Report of a joint FAO/WHO/UNU expert consultation. WHO Technical Report Series 724. WHO, Geneva, pp 113–129

Green AE, Allison RF (1994) Recombination between viral RNA and transgenic plant transcripts. Science 263:1423–1425

Gura T (1999) Biotechnology. New genes boost rice nutrients. Science 285:994–995

Hashimoto W, Momma K, Katsube T, Ohkawa Y, Ishige T, Kito M, Utsumi S, Murata K (1999a) Safety assessment of genetically engineered potatoes with designed soybean glycinin: compositional analyses of the potato tubers and digestibility of the newly expressed protein in transgenic potatoes. J Sci Food Agric 79:1607–1612

Hashimoto W, Momma K, Yoon HJ, Ozawa S, Ishige T, Kito M, Utsumi S, Murata K (1999b) Safety assessment of transgenic potatoes with soybean glycinin by feeding studies in rats. Biosci Biotechnol Biochem 63:1942–1946

Itoh Y (1996) Current status of development of substitute, hypoallergenic and antiallegic foods for food allergy. Bioindustry 13:36–43 (in Japanese)

Jenczewski E, Prosperi JM, Ronfort J (1999) Evidence for gene flow between wild and cultivated *Medicago sativa* (Leguminosae) based on allozyme markers and quantitative traits. Am J Bot 86:677

Katsube T, Kurisaka N, Ogawa M, Maruyama N, Ohtsuka R, Utsumi S, Takaiwa F (1999) Accumulation of soybean glycinin and its assembly with the glutelins in rice. Plant Physiol 120: 1063–1074

Kim CS, Kamiya S, Sato T, Utsumi S, Kito M (1990) Improvement of nutritional value and functional properties of soybean glycinin by protein engineering. Protein Eng 3:725–731

Kito M, Moriyama T, Kimura Y, Kambara H (1993) Changes in plasma lipid levels in young healthy volunteers by adding an extruder-cooked soy protein to conventional meals. Biosci Biotech Biochem 57:354–355

Kota M, Daniell H, Varma S, Garczynski SF, Gould F, Moar WJ (1999) Overexpression of *Bacillus thuringiensis* (Bt) Cry2Aa2 protein in chloroplasts confers resistance to plants against susceptible and Bt-resistant insects. Proc Natl Acad Sci USA 96:1840–1845

Kubo K, Saito M (2000) Current developmental studies and future trends in genetically-modified foods with varying functionality and nutrient contents, and their safety assessment. J Jpn Soc Nutr Food Sci 53:169–174

Lavirk PB, Bartnicki DE, Feldman J, Hammond BG, Keck PJ, Love SL, Naylor MW, Rogan GJ, Sims SR, Fuchs RL (1995) Safety assessment of potatoes resistant to Colorado potato beetle. In: Engel KH, Takeoka GR, Teranishi R (eds) Genetically modified foods. ACS Symposium Series 605. American Chemical Society, Washington, DC, pp 148–158

Losey JE, Rayor LS, Cater ME (1999) Transgenic pollen harms Monarch larvae. Nature 399:214

Moffat AS (1998) Toting up the early harvest of transgenic plants. Science 282:2176–2178

Moffat AS (1999) Crop engineering goes south. Science 285:370–371

Momma K, Hashimoto W, Ozawa S, Kawai S, Katsube T, Takaiwa F, Kito M, Utsumi S, Murata K (1999) Quality and safety evaluation of genetically engineered rice with soybean glycinin: analysis of the grain composition and digestibility of glycinin in transgenic rice. Biosci Biotechnol Biochem 63:314–318

Momma K, Hashimoto W, Yoon HJ, Takaiwa F, Kito M, Utsumi S, Murata K (2000) Safety assessment of rice genetically modified with soybean glycinin by feeding studies on rats. Biosci Biotechnol Biochem 64:1881–1886

Nakamura R, Matsuda T (1994) Gene engineering for hypoallergenic rice. Biosci Industry 52:728–730 (in Japanese)

Nordlee JA, Taylor SL, Townsend JA, Thomas LA, Bush RK (1996) Identification of a Brazil-nut allergen in transgenic soybeans. N Engl J Med 334:688–692

Shelton AM, Roush RT (1999) False reports and the ears of men. Nat Biotechnol 17:832

Smith DB, Roddick JG, Jones JL (1996) Potato glycoalkaloids: some unanswered questions. Trends Food Sci Technol 7:126–131

Sugano M, Goto S, Yamada Y, Yoshida K, Hashimoto Y, Matsuo T, Kimoto M (1990) Cholesterol-lowering activity of various undigested fractions of soybean protein in rats. J Nutr 120: 977–985

Urisu A, Yamada K, Masuda S, Komada H, Wada E, Kondo Y, Horiba F, Tsuruta M, Yasaki T, Yamada M, Torii S, Nakamura R (1991) 16-kilodalton rice protein is one of the major allergens in rice grain extract and responsible for cross-allergenicity between cereal grains in the Poaceae family. Int Arch Allergy Appl Immunol 96:244–252

Utsumi S, Kitagawa S, Katsube T, Higasa T, Kito M, Takaiwa F, Ishige T (1994) Expression and accumulation of normal and modified soybean glycinins in potato tubers. Plant Sci 102: 181–188

Watkinson AR, Freckleton RP, Robinson RA, Sutherland WJ (2000) Prediction of biodiversity response to genetically modified herbicide-tolerant crops. Science 289:1554–1557

Wraight CL, Zangerl AR, Carroll MJ, Brenbaum MR (2000) Absence of toxicity of *Bacillus thuringiensis* pollen to black swallowtails under field conditions. Proc Natl Acad Sci USA 97:7700–7703

10 Chromosomal and Genetic Aberrations in Transgenic Soybean

R.J. SINGH

10.1 Introduction

The soybean [*Glycine max* (L.) Merr.] is an economically important legumi-
nous crop for oil, feed, and soy food products. It contains about 40% protein
and 20% oil in the seed. In the international trade markets, soybean is ranked
number one in oil production (48%) among major oil seed crops (Singh and
Hymowitz 1999). Despite its economic importance, the genetic base of soybean
public cultivars is narrow (Delannay et al. 1983; Gizlice et al. 1993, 1994, 1996;
Salado-Navarro et al. 1993; Sneller 1994; Cui et al. 2000). Soybean breeders have
not yet exploited the wealth of genetic diversity from exotic germplasm, such
as the soybean's ancestor *G. soja* Sieb. and Zucc. or 18 wild perennial species
of the subgenus *Glycine* Willd. Induced-mutation breeding in soybeans has
been used to improve oil quality, tolerance for sulfonylurea herbicides, nitrate-
tolerant symbiotic mutants, and to break the linkage between two closely
linked genes (Singh and Hymowitz 1999).

Conventional breeding has failed to revolutionize gains in soybean yield in
major soybean-producing countries (Singh and Hymowitz 1999). However,
biotechnology is an invaluable tool to broaden the genetic base of soybeans by
overcoming genetic barriers in distant crosses, such as with the tertiary gene
pool (GP-3) of Harlan and de Wet (1971). Soybean somaclonal variants isolated
from tissue cultures in many extensive studies have failed to produce soybeans
with resistance to pests and pathogens and with high yield (Barwale et al. 1986;
Hawbaker et al.1993). Most of the morphological mutants were dwarf, partially
to completely sterile, albino, chlorophyll-deficient, and expressed abnormal
leaf phenotypes (curled, wrinkled, lack of unifoliates), and few of the mutants
exhibited epigenetic inheritance (Graybosch et al. 1987; Shoemaker et al. 1991;
Widholm 1996). It was thus concluded that somaclonal variants are not an
efficient method to improve soybeans. Alien genes of economic importance can
be introgressed into soybean by *Agrobacterium*, particle bombardment, and
electroporation (Finer et al. 1996; Christou 1997). Hinchee et al. (1988) successfully isolated glyphosate-tolerant transformants using the *Agrobacterium*-
mediated gene transfer method, and Padgette et al. (1995) reported on a
transgenic soybean produced by Monsanto under the trade name Round up
Ready. By contrast, Parrott et al. (1989) recorded untransformed soybean plants
by using the same method. Recently, an extremely low frequency (0.03%) of
Agrobacterium-mediated transformants of soybean cv. "Jack" was reported by

Molecular Methods of Plant Analysis, Vol. 22
Testing for Genetic Manipulation in Plants
Edited by J.F. Jackson, H.F. Linskens, and R.B. Inman
© Springer-Verlag Berlin Heidelberg 2002

Yan et al. (2000). Kinney (1996) isolated a stable soybean line with a high oleic acid content (85%) through a particle-bombardment-mediated transformation. Transgenic plants of the soybean have not been produced through electroporation of protoplasts (Christou 1994). Widholm and colleagues claimed transgenic soybeans through electroporation but later retracted their findings (Widholm 1993). Thus, soybean transformation methods are not routinely reproducible (Christou 1997).

During the past decade, progress made in plant transformation has been summarized in several reviews (Fisk and Dandekar 1993; Christou 1994; Klein and Zhang 1994; Vasil 1994; Casas et al. 1995; Songstad et al. 1995; Puddephat et al. 1996; Jouanin et al. 1998; Armstrong 1999). Genetic transformation has improved many economically important crops (soybean, maize, oil seed rapes, cotton) grown commercially (Dunwell 2000). Embryogenic suspension cultures of the soybean can be initiated from globular-stage somatic embryos arising on immature cotyledons (Finer and Nagasawa 1988). Genes can be delivered by *Agrobacterium*-mediated transformation of cotyledonary explants (Hinchee et al. 1988; Clemente et al. 2000) and somatic embryos from immature cotyledonary explants (Parrott et al. 1989; Stewart et al. 1996; Santarém et al. 1998) or to the embryogenic suspension cultures by microprojectile particle bombardment, invented by Klein et al. (1987), (McCabe et al. 1988; Finer and McMullen 1991). The first (primary) soybean transformants are often sterile and display a range of phenotypic variability. These abnormalities are attributed mostly to chromosomal aberrations induced by the culture conditions (Singh et al. 1998).

Chromosomal aberrations (numerical and structural changes) are a common occurrence in cell and tissue culture-derived calluses and their regenerants (Singh 1993). Chromosomally abnormal soybean cultures used in transformation experiments regenerate mostly sterile plants (Singh et al. 1998). Soybean is considered one of the most difficult large-seeded legumes from which to regenerate plants through cell and tissue cultures. Earlier reports suggested that a reproducible method of regeneration depends upon the type and age of explants, media composition (basic salts and growth hormone combinations) and genotypes (Christianson et al. 1983; Barwale et al. 1986; Lazzeri et al. 1987a, b; Wright et al. 1987; Bailey et al. 1993). The degree of sterile transgenic soybean plants is related to the period the culture is nurtured on 2,4-D prior to the transformation experiments, post-transformation stresses, and the genetic background of the explants. The objective of this article is to discuss the cytological characterization of transgenic soybeans.

10.2 Times in Culture with 2,4-D Prior to Transformation

Generally, older cultures lose the capacity to regenerate plants because they either carry high ploidy in the tissues or accumulate an increasing number of aneuploid cells that lead to the loss of a balanced chromosomal constitution of the cells (Singh 1993). Thus, these cultures, when used in the transformation experiments, will generate a low frequency of morphologically deformed plants with unbalanced chromosome constitution. Singh et al. (1998) examined chromosome counts (mitotic pro-metaphase and metaphase) in embryogenic suspensions cultures (non-transgenic) and in roots from developing somatic embryos (either transgenic or non-transgenic) from T_0, T_1, and T_2 transgenic Asgrow genotypes A2242 and A2872.

Seeds of A2242 (control) carried the expected $2n = 40$ normal chromosomes. The chromosome counts of suspension cultures and germinating somatic embryos from culture 817 are extremely informative (Table 10.1). Tetraploidy occurred even in the germinating somatic embryos from 7.83 months on 2,4-D; three samples showed $2n = 80$ normal chromosomes and two samples, in addition to tetraploidy, carried $2n = 79 + 1$ dicentric chromosomes.

Table 10.1. Chromosome analysis at somatic metaphase in the transgenic Asgrow soybean genotype A2242. (Singh et al. 1998)

Culture ID	R0 phenotype	Origin of roots	Months on 2,4-D	No. samples	2n	Karyotype
A2242		Control		5	40	Normal
22–1	Diploid	T1	6.43	7	40	Normal
22–1	Tetraploid	T1	6.90	5	40;41	39 + (1); 38 + (3)[a]
22–1	Diploid	T1	6.90	2	40	Normal
22–1	Diploid	T1	6.96	9	40	Normal
22–1	Diploid	T1	6.96	2	80	Normal
22–1	Diploid	T1	7.00	8	40	Normal
22–1	Diploid	T1	7.30	9	80	Normal
22–1	Diploid	T1	7.96	8	80	Normal
22–1	Diploid	T1	9.00	1	80	Normal
22–1	Tetraploid	T0	11.47	3	80	Normal
22–1	Tetraploid	T0	15.36	1	80	Normal
22–1	–	Embryo	15.36	5	80	3, 80; 2, 79 + 1[b]
22–1	–	Embryo	16.73	4	80	3, 80; 1, 40 + 80[c]
817	–	Embryo	7.17	4	40	3, 40; 1, 39 + 1[d]
817	–	Embryo	7.33	5	40	4, 40; 1, 39 + 1[d]
817	–	Embryo	7.83	5	80	3, 80; 2, 79 + 1[e]
817	–	Embryo	8.70	1	40	Normal
817	–	T0	9.13	1	40	Normal
817	–	Suspension	11.26	1	80	Normal
817	–	Embryo	12.43	2	80	Normal
825	–	Suspension	6.86	1	40	Normal
828	–	Embryo	2.86	1	40	Normal
826	–	Suspension	4.20	1	80	Normal

[a] Three small metacentric chromosomes.
[b] One megachromosome; chimaera 40 + 80 chromosomes.
[d] Long chromosome.
[e] One sample with 79 + 1 dicentric chromosomes and another sample with 79 + 1 fused centromeric chromosomes.

Cells in 11.26 months on 2,4-D suspensions cultures showed 2n = 80 chromosomes and transgenic plants recovered from this culture were tetraploid. Embryo suspensions of culture 826, which were nurtured for 4.20 months on 2,4-D, displayed cells with 2n = 80 chromosomes. Suspension culture 825 (6.86 months on 2,4-D) and somatic embryo culture 828 (only 2.86 months on 2,4-D) showed diploid cells. The germinating embryos derived from cultures ranging from 7.17 to 16.73 months on 2,4-D possessed unbalanced chromosomes (Table 10.1). The older cell lines were not desirable for transformation because chromosomal and genetic abnormalities occur with the age. This may lead to difficulty in plant recovery, and if plants are regenerated seed sterility is frequently recorded.

In order to ensure morphologically normal transformants with complete fertility, Stewart et al. (1996) bombarded 3-month-old globular-stage embryos of soybean cv. "Jack" and used post-bombardment selection of transgenic lines on a solidified medium instead of liquid medium. The solid medium promoted slower growth of transgenic cells and slower death of nontransgenic cells. Transgenic plants were not analyzed cytologically but all plants were fertile and morphologically normal. Maughan et al. (1999) isolated four normal and fertile β-casein-transgenic soybean cv. "Jack" plants via particle bombardment of 3- to 4-month-old somatic embryos nurtured on solid medium. Hazel et al. (1998) examined the growth characteristics and transformability of embryogenic cultures of soybean cv. "Jack" and Asgrow A2872. The most transformable cultures comprised tightly packed globular structures and cytoplasmic-rich cells in the outermost layers of the tissues with the highest mitotic index. By contrast, the outer layers of less transformable cultures had more lobed cells with prominent vacuoles. Santarém and Finer (1999) used 4-week-old proliferative embryogenic tissue of soybean cv. "Jack" maintained on semi-solid medium for transformation by particle bombardment. They produced fertile transgenic soybeans 11–12 months following culture initiation. Santarém and Finer (1999) also claimed a significant improvement over bombardment of embryogenic liquid suspension culture tissue of soybean in which transformability was recorded at least 6 months after culture initiation (Hazel et al. 1998). Clemente et al. (2000) produced 156 primary transformants by Agrobacterium-mediated transformation. Glyphosate-tolerant shoots were identified after 2–3 months of selection. The R_0 (T_0) plants were fertile (262 seeds per plant) and did not express gross phenotypic abnormalities. This suggests that transformation in the soybean must be conducted using young (1–3 months) cultures.

The plants regenerated from the cultures, listed in Table 10.1, also inherited chromosomal abnormalities. Chromosomes of nine T_1 populations from culture 22–1 were examined. The parental cultures were in contact with 2,4-D for 6.43–9.00 months prior to transformation. Although T_0 plants from eight cultures expressed normal diploid morphological features, T_1 progenies from four populations carried 2n = 80 chromosomes and four populations had 2n = 40 chromosomes (Fig. 10.1A). The T_0 phenotype of one T_1 population of culture 22–1 was similar to the tetraploid. These plants showed dark-green leathery leaves and produced mostly one-seeded pods. Chromosome counts from five T_1 seedlings showed 2n = 39 + 1 small metacentric chromosome (Fig. 10.1B) in three plants, and one plant each contained 2n = 38 + 3 small metacentric chromosomes (Fig. 10.1C) and $2n$ = 40 chromosomes. The 40-

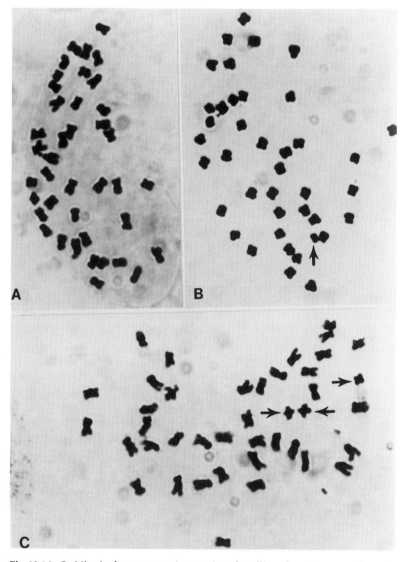

Fig. 10.1A–C. Mitotic chromosomes in root tips of seedlings from Asgrow soybean line A2242 R_1 generation. A 2n = 40 showing normal karyotype, B 2n = 39 + 1 small metacentric chromosomes (*arrow*), C 2n = 38 + 3 small metacentric chromosomes (*arrows*). (Singh et al. 1998)

chromosome plant may have had a small deletion, which, however, could not be detected cytologically, or may have carried desynaptic or asynaptic genes. Four T_0-derived plants from culture 22–1 (3 plants from 11.47 months on 2,4-D and one plant from 15.36 months on 2,4-D) were morphologically tetraploid and this was confirmed cytologically as all plants carried 2n = 80 chromosomes. Three transformants had 2n = 80 (Fig. 10.2A) and two contained 79 + 1 megachromosomes (Fig. 10.2B). Most reports agree that chromosome structural and numerical changes in culture are induced by 2, 4-D (Singh 1993). Thus, it was suggested to cytologically examine the chromosome stability of pre-transformed material to ensure that it is devoid of such changes (Poulsen 1996).

It has been established that various stresses, such as medium composition, age of culture, and the nature of the culture, i.e. morphogenic vs. non-

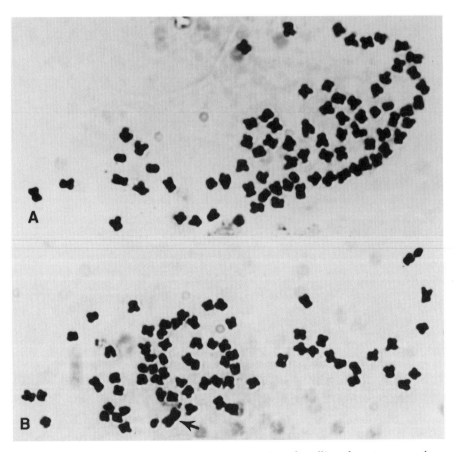

Fig. 10.2A,B. Mitotic metaphase chromosomes in root tips of seedlings from Asgrow soybean line A2242 grown in agar. **A** 2n = 80 showing normal karyotype, **B** 2n = 79 + 1 mega (monocentric) chromosome (*arrow*). (Singh et al. 1998)

morphogenic, genetic backgrounds of the explants, kind of media, and in vitro culture during the passage of plant regeneration, induce chromosomal instability (Singh 1993; D'Amato 1995). Choi et al. (2000) studied chromosome constitution in transgenic vs. nontransgenic barley (2n = 14). They concluded that the extent of ploidy changes in transgenic plants was intensified, perhaps due to the additional stresses that occurred during transformation. The delivery of foreign DNA into plant cells involves several stressful events, such as vacuuming, cellular damage by microprojectile bombardment, the selection process, and the growth of dying transformed cells for a prolonged time during recovery, that cause chromosomal aberrations. Chromosomal aberrations are attributed to spindle failure by growth hormones that causes endoreduplication, c-mitosis, nuclear fragmentation (amitosis), multipolar configurations, and lagging chromosomes (Singh 1993).

10.3 Genetic Background of the Explants

Chromosomal aberrations induced during culture are genotype-dependent (Singh 1993). Singh et al. (1998) observed that tissues and primary transformants from soybean genotype A2872 did not display chromosomal abnormalities (Table 10.2), while soybean genotype A2242 was highly unstable cytologically (Table 10.1). Nine selfed seeds from A2872 (control) showed 2n = 40 chromosomes with normal karyotype. Culture age on 2,4-D ranged from 8.26 to 32.30 months. Embryo suspensions, germinating embryos, and T_0 and T_1 plants expressed normal diploid-like phenotypes and, as expected, all plants carried 2n = 40 chromosomes. However, five T_1 plants from a 29.43-month-old culture of 5-2 had tetraploid morphological features with one-seeded pods, but showed 2n = 40 chromosomes (Table 10.2). The abnormalities may be genic (desynaptic or asynaptic, male sterile and female fertile). These morphotypes often express tetraploid phenotypes (slow growth, dark-green leathery leaves, clustered flowers, empty pods) with partial-to-complete sterility and carrying one seed per pod (Palmer and Kilen 1987). Interestingly, the primary transformant from A2872 contained 2n = 40 chromosomes though the culture was on 2,4-D for 32.30 months. This elucidates that a genotype may be highly responsive to culture conditions but may be prone to chromosome aberrations (Hermsen 1994).

Aragão et al. (2000) formulated a soybean transformation procedure that produced a high frequency of fertile transgenic soybeans and was variety-independent. Their technique includes microparticle bombardment (*ahas* gene: a selectable marker gene isolated from *Arabidopsis thaliana* that contains a mutation at position 653 bp) of the soybean meristematic region, culturing in selection medium with imazapyr herbicides followed by a multiple shooting induction. They claim a 200-fold increase in the recovery of transgenic soybean plants over the methods of Christou (1997). An experiment to enhance rapid

Table 10.2. Chromosome analysis at somatic metaphase in the transgenic Asgrow soybean genotype A2872. (Singh et al. 1998)

Culture ID	R0 phenotype	Origin of roots	Months on 2,4-D	No. samples	2n	Karyotype
A2872		Control		9	40	Normal
821	–	Suspension	8.26	1	40	Normal
821	–	Embryo	9.23	5	40	Normal
2–1	Diploid	T1	22.76	1	40	Normal
2–2	Diploid	T1	11.20	1	40	Normal
2–2	Diploid	T1	18.03	1	40	Normal
2.4	Diploid	T1	13.16	1	40	Normal
2–4	Diploid	T1	20.00	1	40	Normal
5–1	Diploid	T1	20.50	2	40	Normal
5–1	Diploid	T1	21.50	2	40	Normal
5–2	Diploid	T1	19.06	4	40	Normal
5–2	Diploid	T1	20.20	1	40	Normal
5–2	Diploid	T1	20.73	1	40	Normal
5–2	Diploid	T1	21.20	2	40	Normal
5–2	Diploid	T1	29.00	3	40	Normal
5–2	Diploid	T1	29.43	3	40	Normal
5–2	Tetraploid	T1	29.43	5	40	Normal
90–2	Diploid	T1	19.06	1	40	Normal
90–2	Diploid	T1	32.30	4	40	Normal
801	–	Embryo	9.23	2	40	Normal

production of stable, normal, and fertile transformants would be warranted to
formulate a protocol from the procedures of Santarém and Finer (1999) and
Aragão et al. (2000).

10.4 Seed Fertility in Transgenic Soybean

Morphological variants, particularly seed sterility in transgenic crops, have
often been recorded (Christey and Sinclair 1992; Conner et al. 1994; Ghosh
Biswas et al. 1994; Austin et al. 1995; El-Kharbotly et al. 1995; Fütterer and
Potrykus 1995; Lynch et al. 1995; Schulze et al. 1995; Shewry et al. 1995;
Widholm 1996; Hadi et al. 1996; Liu et al.1996). *Agrobacterium*-mediated transformation produced diploid (2n = 24) (El-Kharbotly et al. 1995) and tetraploid
(2n = 2x = 48) potato (Conner et al. 1994). The diploid transgenic potato that
conferred resistance to *Phytophthora infestans* inherited reduced male fertility while female fertility was less markedly influenced. All transformed
tetraploid potatoes contained a normal chromosome complement but morphological changes including low yield and small tubers were observed in the
field-grown plants compared to control.

Agrobacterium-mediated transformation has been extremely effective in dicotyledonous crops. However, Ishida et al. (1996) developed an *Agrobacterium*-mediated transformation protocol for the high-efficiency transformation of maize. They produced 120 transgenic maize plants, not examined cytologically, and almost all of the plants were morphologically normal; 70% of them produced normal seeds.

Christey and Sinclair (1992) obtained kale, rape and turnip transformants through *Agrobacterium rhizogenes*-mediated transformation. Transgenic plants were successfully produced from hairy roots of all cultivars. Morphological changes with an increase in leaf edge serration, leaf wrinkling and plagiotropic roots were observed in some plants while in other lines phenotypic alterations were barely noticeable. Transgenic plants were not examined cytologically.

Toriyama et al. (1988) produced five transgenic rice plants after direct gene transfer into protoplasts through electroporation-mediated transformation. One plant was diploid ($2n = 24$), three plants were triploid ($2n = 3x = 36$) and one plant was unidentified. Ghosh-Biswas et al. (1994) produced 73 transgenic rice cv. IR 43 plants by direct transfer of genes to protoplasts; 29 plants reached maturity in the greenhouse. Eleven plants flowered but did not produce seed. However, two protoplast-derived nontransgenic plants set seeds. Protoplast-derived plants (transgenic and nontransgenic) had fewer tillers, narrower leaves and were shorter than seed-derived plants. Lynch et al. (1995) observed that electroporation-mediated transgenic rice ($2n = 24$) plants were shorter, took longer to flower and showed partial sterility when compared to nontransgenic plants. Fertile transgenic rice plants have been produced through electroporation-mediated transformation (Shimamoto et al.1989; Xu and Li 1994).

Another method to insert a foreign gene into the protoplast is by polyethylene glycol-mediated DNA transformation. Hall et al. (1996) produced a high-efficiency procedure for the generation of transgenic sugar beets from stomatal guard cells. They examined ploidy level by flow cytometry and recorded 75% diploid transformants. Seed production was normal and the average frequency of germination was 96%. Lin et al. (1995) transferred 61 rice plants (via polyethylene-glycol-mediated transformation) to a greenhouse and 28 were fertile. Seed set per plant ranged from 10 to 260. The cause of seed sterility was not determined.

Primary transformants (T_0) in soybean genotype A2242 exhibited a range of fertility that depended upon culture and duration of time tissue spent on 2,4-D (Table 10.3). The chromosome counts of ten seeds from randomly selected T_0 plants revealed normal ($2n = 40$) chromosomes from the young cultures regardless of seed set. For example, ten seeds from six T_0 plants examined cytologically were 3.20 months on 2,4-D with seed set ranging from 0 to 363. All plants carried $2n = 40$ chromosomes. Tetraploidy ($2n = 80$) and aneuploidy (near diploidy) predominated in plants obtained from the older cultures (Table 10.3). Culture 828–2 produced 12 T_0 plants from 11 months on 2,4-D.

Table 10.3. Culture identification (ID), plant ID, months in culture with 2,4-D before transformation, total number of transformants recovered, seed set range[a], and chromosome number in selected primary transformants in soybean cultivar Asgrow 2242

Culture ID	Plant ID	Months 2,4-D	Total number plants	Seed set range	2n		
					40	80	41
828-1	671-3-4	3.20	6	0-363	14	–	–
828-1	668-8-1	5.50	4	5-50	9	–	–
828-1	668-1-3	5.60	9	14-219	10	–	–
828-1	668-1-6	5.60	5	6-65	9	–	1
828-1	668-1-5	5.60	2	15-24	7	–	3
828-1	668-1-8	5.60	2	85-101	7	–	–
828-1	668-1-12	5.60	9	75-185	18	–	–
828-1	668-1-13	5.60	5	59-178	10	–	–
828-1	668-2-3	5.60	10	6-53	–	20	–
828-1	668-2-7	5.60	7	6-111	–	10	–
828-1	668-4-3	5.60	6	2-12	4	–	–
828-1	668-4-15	5.60	5	8-51	1	–	–
828-1	668-1-1	6.00	4	0-225	4	–	–
22-1	549-2-2	6.43	1	25	3	–	1
22-1	549-4-6	6.43	2	63-206	7	–	3
22-1	549-4-10	6.43	5	90-321	20	–	–
22-1	557-1-6	7.30	3	9-65	–	10	–
22-1	557-2-3	7.30	7	90-321	–	15	–
817	610-5-1	7.33	10	10-495	54	–	–
817	610-6-1	7.83	5	56-190	23	–	14+1[a]
817	610-7-1	7.83	10	52-329	–	4	–
22-1	557-2-7	7.96	10	33-350	–	30	–
22-1	557-5-2	7.96	9	65-210	–	9	–
22-1	566-5-1	9.50	1	27	–	10	–
828-2	677-3-1	11.00	12	50-268	5	5	–
817	647-1-1	12.43	5	0-19	5	–	–
817	647-2-1	12.43	5	25-66	–	3	–
817	647-15-7	12.43	5	30-133	–	10	–
817	647-15-4	12.43	2	7-36	10	–	–
817	652-6-1	13.30	12	0-235	–	9	–
828-1	668-7-2	13.80	2	28-38	4	–	2
22-1	587-5-1	15.36	2	7-18	–	2	–
22-1	587-3-1	15.36	3	4-8	–	6	–
22-1	609-6-1	16.00	1	9	–	3	–

[a] T.M. Klein personal communication.

Seeds of ten plants germinated showed five diploid and five tetraploid plants. Seed fertility in nine plants was more than 101 seeds per plant while seed set in three plants ranged from 21 to 100 per plant. Occasionally, T_0 plants produced low seeds (2–12; 0–19; 7–36) and were diploid. Thus, sterility may be attributed either to culture conditions (epigenetic) or desynaptic and asynaptic gene mutations or minor chromosomal deletions which cannot be detected cytologically. Schulze et al. (1995) observed fruit development after selfing transgenic cucumber, however, none of the harvested fruits contained seeds. Simmonds and Donaldson (2000) produced via particle bombardment fertile transgenic soybeans from young proliferative cultures, while sterile plants were recovered from 12- to 14-month-old cultures. The cause of sterility was not established. This study suggests that transformation experiments conducted on young cultures produce completely fertile transformants without chromosomal aberrations.

10.5 Cytological Basis of Gene Silencing

The loss or low expression of foreign genes and their unexpected segregation are routinely observed in transformed crops (Chupeau et al. 1989; Fromm et al. 1990; Somers et al. 1994; Fütterer and Potrykus 1995; Meyer 1995; Senior 1998; Kooter et al. 1999). Somers et al. (1994) examined GUS activity in R_1 (T_1) and some R_2 (T_2) and R_3 (T_3) generations in 15 transgenic families of oat (2n = 6x = 42). Six families expressed an aberrant segregation ratio, seven families segregated in a 3:1 GUS:GUS‾ratio, and two families segregated 15:1 for GUS activity. Foreign genes in transgenic common bean plants produced from particle-bombardment were not expressed in 12 (44%) plants and two plants showed poor transmission of the insert gene (1:10) in the R_1 (T_1) generation, although all plants had a normal phenotype and were fertile (Aragão et al. 1996). This may be due chimerism in T_0 plants.

An extensive review of genetic transformation research and gene expression in the Poaceae by Fütterer and Potrykus (1995) revealed that transgene expressions in the progeny of transgenic plants is quite unpredictable. The departure from Mendelian inheritance occurs in transformants when inserts are located in the extrachromosomal DNA (mitochondrial or chloroplast genomes). Several genomic factors, such as incorrect crossing-over during meiosis, spontaneous or induced mutations, ploidy, aneuploidy, sex chromosomes and transposable elements, cause deviations in the Mendelian inheritance of the transgenes (Maessen 1997).

In transformation experiments, genes may be physically present but gene activity may be poorly expressed or totally lost in subsequent generations. This phenomenon is known as co-suppression or gene silencing (Matzke and Matzke 1995; Matzke et al. 2000; Stam et al. 1997). In co-suppression, foreign genes (transgenes) cause the silencing of endogenous plant genes if they are sufficiently homologous (Stam et al. 1997). An excellent example of co-suppression was shown in tobacco by Brandle et al. (1995). They produced a transgenic tobacco line carrying the mutant A. thaliana acetohydroxyacid synthase gene csr1-1 that expressed a high level of resistance to the sulfonylurea herbicide chlorsulfuron. The instability of herbicide resistance was observed during subsequent field trials and was not anticipated from the initial greenhouse screening. Hemizygous plants from this line were resistant but 59% of the homozygous plants were damaged by the herbicide. Damage was correlated with co-suppression of the *csr1–1* transgene and the endogenous tobacco AHAS genes, *sur A* and *sur B*. Differences in glyphosate tolerance in sugar-beet transformants between greenhouse and the field were also reported by Mannerlöf et al. (1997). The disparity was attributed to differences between the environments, variation in gene expression caused by copy number, position or methylation effects. Based on this information, the authors suggested that co-suppression was triggered by agro-climatic conditions and that the initial greenhouse study was not predictable. This study demonstrates that the stability of inserts in a crop should be field-tested under a wide range of agro-eco-geo-climatic growing conditions (day length, drought, heat, irregular weather) for several years before the cultivar is released commercially.

Singh et al. (1998) presented a cytological clue that may help explain aberrant segregation ratios or loss of transgene sequences, and which may be applicable in some cases. For example, the selfed population of a plant with $2n = 39 + 1$ metacentric chromosomes identified in soybean genotype A2242 was expected to segregate plants in a ratio of 1 ($2n = 40$):2 ($2n = 39 + 1$ metacentric):1 ($2n = 38 = 2$ metacentrics). Diploid plants will be normal and fertile and may not express the introgressed genes if this gene is in the deleted chromosomes. Matzke et al. (1994) attributed an erratic inheritance in a transgenic tobacco line to aneuploidy ($2n = 49$ or 50).

The stability of transformants is associated with the insertion of transgene (s) in the regions of chromosomes. According to Matzke et al. (2000), different regions of the genomes vary in their ability to tolerate foreign DNA, resulting in erratic expression. Iglesias et al. (1997) demonstrated by fluorescence in situ hybridization (FISH) that two stably expressed inserts in tobacco were present in the vicinity of telomeres, while two unstably expressed inserts were located at intercalary and paracentromeric regions. It is worthwhile to examine transgenic soybean line by FISH in order to determine the precise location of transgenes. Based on pachytene chromosome analysis, Singh and Hymowitz (1988) established that centromeric regions of the soybean chromosomes are heterochromatic and distal segments are euchromatic. A FISH study is warranted using either primary trisomics or tetrasomics to accurately locate several stable soybean transgenes.

10.6 Conclusions

Information on chromosomal and genetic aberrations in transgenic crops is lacking. An early chromosome count (pre transformation) of cells, callus, and embryo suspension will ensure insertion of foreign genes into the normal chromosome complement of the plants. This will result in production of morphologically normal, fertile and genetically stable transgenic crops. The rejection of chromosomally abnormal cultures prior to the transformation process is cost-effective as it will save time, labor and eventually frustration.

Acknowledgements. The author thanks Dr. Theodore Hymowitz for reading the manuscript, and Dr. Ted Klein for providing seed fertility data on the transgenic soybean.

References

Aragão FJL, Barros LMG, Brasileiro ACM, Ribeiro SG, Smith FD, Sanford JC, Faria JC, Rech EL (1996) Inheritance of foreign genes in transgenic bean (*Phaseolus vulgaris* L.). Co-transformed via particle bombardment. Theor Appl Genet 93:142–150

Aragão FJL, Sarokin L, Vianna GR, Rech EL (2000) Selection of transgenic meristematic cells utilizing a herbicidal molecule results in the recovery of fertile transgenic soybean [*Glycine max* (L.) Merril] plants at a high frequency. Theor Appl Genet 101:1–6

Armstrong CL (1999) The first decade of maize transformation: a review and future perspective. Maydica 44:101–109

Austin S, Bingham ET, Mathews DE, Shahan MN, Will J, Burgess RR (1995) Production and field performance of transgenic alfalfa (*Medicago sativa* L.) expressing alpha-amylase and manganese-dependent lignin peroxidase. Euphytica 85:381–393

Bailey MA, Boerma HR, Parrott WA (1993) Genotype – specific optimization of plant regeneration from somatic embryos of soybean. Plant Sci 93:117–120

Barwale UB, Meyer MM Jr, Widholm JM (1986) Screening of *Glycine max* and *Glycine soja* genotypes for multiple shoot formation at the cotyledonary node. Theor Appl Genet 72:423–428

Brandle JE, McHugh SG, James L, Labbé H, Miki BL (1995) Instability of transgene expression in field grown tobacco carrying the *csr1-1* gene for sulfonylurea herbicide resistance. Bio/Technology 13:994–998

Casas AM, Kononowicz AK, Bressan RA, Hasegawa PM (1995) Cereal transformation through particle bombardment. Plant Breed Rev 13:235–264

Choi HW, Lemaux PG, Cho MJ (2000) Increased chromosomal variation in transgenic versus non-transgenic barley (*Hordeum vulgare* L.) plants. Crop Sci 40:524–533

Christey MC, Sinclair BK (1992) Regeneration of transgenic kale (*Brassica oleracea* var. *acephala*), rape (*B. napus*) and turnip (*B. campestris* var. *rapifera*) plants via *Agrobacterium rhizogenes* mediated transformation. Plant Sci 87:161–169

Christianson ML, Warnick DA, Carlson PS (1983) A morphogenetically competent soybean suspension culture. Science 222:632–634

Christou P (1994) The biotechnology of crop legumes. Euphytica 74:165–185

Christou P (1997) Biotechnology applied to grain legumes. Field Crops Res 53:83–97

Chupeau M, Bellini C, Guerche P, Maisonneuve B, Vastra G, Chupeau Y (1989) Transgenic plants of lettuce (*Lactuca sativa*) obtained through electroporation of protoplasts. Bio/Technology 7:503–508

Clemente TE, LaVallee BJ, Howe AR, Conner-Ward D, Rozman RJ, Hunter PE, Broyles DL, Kasten DS, Hinchee MA (2000) Progeny analysis of glyphosate selected transgenic soybeans derived from *Agrobacterium*-mediated transformation. Crop Sci 40:797–803

Conner AJ, Williams MK, Abernethy DJ, Fletcher PJ, Genet RA (1994) Field performance of transgenic potatoes. N Z J Crop Hortic Sci 22:361–371

Cui Z, Carter TE Jr, Burton JW (2000) Genetic base of 651 Chinese soybean cultivars released during 1923 to 1995. Crop Sci 40:1470–1481

D'Amato F (1995) Aneusomaty in vivo and in vitro in higher plants. Caryologia 48:85–103

Delannay X, Rodgers DM, Palmer RG (1983) Relative genetic contributions among ancestral lines to North American soybean cultivars. Crop Sci 23:944–949

Dunwell JM (2000) Transgenic approaches to crop improvement. J Exp Bot 51:487–496

El-Kharbotly A, Jacobsen E, Stiekema WJ, Pereira A (1995) Genetic localisation of transformation competence in diploid potato. Theor Appl Genet 91:557–562

Finer JJ, McMullen MD (1991) Transformation of soybean via particle bombardment of embryogenic suspension culture tissue. In Vitro Cell Dev Biol 27P:175–182

Finer JJ, Nagasawa A (1988) Development of an embryonic suspension culture of soybean (*Glycine max* Merrill). Plant Cell Tissue Organ Cult 15:125–136

Finer JJ, Cheng T-S, Verma DPS (1996) Soybean transformation: technologies and progress. In: Verma DPS, Shoemaker RC (eds) Soybean genetics, molecular biology and biotechnology. CAB International, Wallingford, UK, pp 249–262

Fisk HJ, Dandekar AM (1993) The introduction and expression of transgenes in plants. Sci Hortic 55:5–36

Fromm ME, Morrish F, Armstrong C, Williams R, Thomas J, Klein TM (1990) Inheritance and expression of chimeric genes in the progeny of transgenic maize plants. Bio/Technology 8:833–839

Fütterer J, Potrykus I (1995) Transformation of Poaceae and gene expression in transgenic plants. Agronomie 15:309–319

Ghosh Biswas GC, Iglesias VA, Datta SK, Potrykus I (1994) Transgenic Indica rice (*Oryza sativa* L.) plants obtained by direct gene transfer to protoplasts. J Biotechnol 32:1–10

Gizlice Z, Carter TE Jr, Burton JW (1993) Genetic diversity in North American soybean: I. Multivariate analysis of founding stock and relation to coefficient of parentage. Crop Sci 33:614–620

Gizlice Z, Carter TE Jr, Burton JW (1994) Genetic basis for North American public soybean cultivars released between 1947 and 1988. Crop Sci 34:1143–1151

Gizlice Z, Carter TE Jr, Gerig TM, Burton JW (1996) Genetic diversity patterns in North American public soybean cultivars based on coefficient of parentage. Crop Sci 36:753–765

Graybosch RA, Edge ME, Delannay X (1987) Somaclonal variation in soybean plants regenerated from the cotyledonary node tissue culture system. Crop Sci 27:803–806

Hadi MZ, McMullen MD, Finer JJ (1996) Transformation of 12 different plasmids into soybean via particle bombardment. Plant Cell Rep 15:500–505

Hall RD, Riksen-Bruinsma T, Weyens GJ, Rosquin IJ, Denys PN, Evans IJ, Lathouwers JE, Lefèbvre MP, Dunwell JM, van Tunen A, Krens FA (1996) A high frequency technique for the generation of transgenic sugar beets from stomatal guard cells. Nat Biotechnol 14:1133–1138

Harlan JR, de Wet JMJ (1971) Toward a rational classification of cultivated plants. Taxon 20: 509–517

Hazel CB, Klein TM, Anis M, Wilde HD, Parrott WA (1998) Growth characteristics and transformability of soybean embryogenic cultures. Plant Cell Rep 17:765–772

Hawbaker MS, Fehr WR, Mansur LM, Shoemaker RC, Palmer RG (1993) Genetic variation for quantitative traits in soybean lines derived from tissue culture. Theor Appl Genet 87:49–53

Hermsen JGTH (1994) Introgression of genes from wild species, including molecular and cellular approaches. In: Bradshaw JE, MacKay GR (eds) Potato genetics. CAB International, Wallingford, UK, pp 515–538

Hinchee MAW, Connor-Ward DV, Newell CA, McDonnell RE, Sato SJ, Gasser CS, Fischhoff DA, Re DB, Fraley RT, Horsch RB (1988) Production of transgenic soybean plants using *Agrobacterium*-mediated DNA transfer. Bio/Technology 6:915–922

Iglesias VA, Moscone EA, Papp I, Neuhuber F, Michalowski S, Phelan T, Spiker S, Matzke M, Matzke AJM (1997) Molecular and cytogenetic analysis of stably and unstably expressed transgene loci in tobacco. Plant Cell 9:1251–1264

Ishida Y, Saito H, Ohta S, Hiei Y, Komari T, Kumashiro T (1996) High efficiency transformation of maize (*Zea mays* L.) mediated by *Agrobacterium tumefaciens*. Nat Biotechnol 14:745–750

Jouanin L, Bondé-Bottino M, Girard C, Morrot G, Giband M (1998) Transgenic plants for insect resistance. Plant Sci 131:1–11

Kinney AJ (1996) Development of genetically engineered soybean oils for food applications. J Food Lipids 3:273–292

Klein TM, Zhang W (1994) Progress in the genetic transformation of recalcitrant crop species. Aspects Appl Biol 39:35–44

Klein TM, Wolf ED, Wu R, Sanford JC (1987) High-velocity microprojectiles for delivering nucleic acids into living cells. Nature 327:70–73

Kooter JM, Matzke MA, Meyer P (1999) Listening to the silent genes: transgene silencing, gene regulation and pathogen control. Trends Plant Sci 9:340–347

Lazzeri PA, Hildebrand DF, Collins GB (1987a) Soybean somatic embryogenesis: effects of hormones and culture manipulations. Plant Cell Tissue Organ Cult 10:197–208

Lazzeri PA, Hildebrand DF, Collins GB (1987b) Soybean somatic embryogenesis: effects of nutritional, physical and chemical factors. Plant Cell Tissue Organ Cult 10:209–220

Lin W, Anuratha CS, Datta K, Potrykus I, Muthukrishnan S, Datta SK (1995) Genetic engineering of rice for resistance to sheath blight. Bio/Technology 13:686–691

Liu W, Torisky RS, McAllister KP, Avdiushko S, Hildebrand D, Collins GB (1996) Somatic embryo cycling: evaluation of a novel transformation and assay system for seed-specific gene expression in soybean. Plant Cell Tissue Organ Cult 47:33–42

Lynch PT, Jones J, Blackhall NW, Davey MR, Power JB, Cocking EC, Nelson MR, Bigelow DM, Orum TV, Orth CE, Schuh W (1995) The phenotypic characterisation of R_2 generation transgenic rice plants under field and glasshouse conditions. Euphytica 85:395–401

Maessen GDF (1997) Genomic stability and stability of expression in genetically modified plants. Acta Bot Neerl 46:3–24

Mannerlöf M, Tuvesson S, Steen P, Tenning P (1997) Transgenic sugar beet tolerant to glyphosate. Euphytica 94:83–91

Matzke AJ, Matzke MA (1995) Trans-inactivation of homologous sequences in Nicotiana tabacum. Current Top Microbiol Immunol 197:1–14

Matzke MA, Moscone EA, Park Y-D, Papp I, Oberkofler H, Neuhuber F, Matzke AJM (1994) Inheritance and expression of a transgene insert in an aneuploid tobacco line. Mol Gen Genet 245:471–485

Matzke MA, Mette MF, Kunz C, Jakowitsch J, Matzke AJM (2000) Homology-dependent gene silencing in transgenic plants: links to cellular defense responses and genome evolution. In: Gustafson JP (ed) Genomes. Kluwer/Plenum, New York, pp 141–162

Maughan PJ, Philip R, Cho MJ, Widholm JM, Vodkin LO (1999) Biolistic transformation, expression, and inheritance of bovine β-casein in soybean (Glycine max). In Vitro Cell Dev Biol-Plant 35:344–349

McCabe DE, Swain WF, Martinell BJ, Christou P (1988) Stable transformation of soybean (Glycine max) by particle acceleration. Bio/Technology 6:923–926

Meyer P (1995) Variation of transgene expression in plants. Euphytica 85:359–366

Padgette SR, Kolacz KH, Delannay X, Re DB, La Vallee BJ, Tinius CN, Rhodes WK, Otero YI, Barry GF, Eichholtz DA, Peschke VM, Nida DL, Taylor NB, Kishore GM (1995) Development, identification, and characterization of glyphosate-tolerant soybean line. Crop Sci 35:1451–1461

Palmer RG, Kilen TC (1987) Qualitative genetics and cytogenetics. In: Wilcox JR (ed) Soybeans: improvement, production, and uses, 2nd edn. Agronomy Monogr No 16, ASA-CSSA-SSSA, Madison, WI, pp 135–209

Parrott WA, Hoffman LM, Hildebrand DF, Williams EG, Collins GB (1989) Recovery of primary transformants of soybean. Plant Cell Rep 7:615–617

Poulsen GB (1996) Genetic transformation of Brassica. Plant Breed 115:209–225

Puddephat IJ, Riggs TJ, Fenning TM (1996) Transformation of Brassica oleracea L.: a critical review. Mol Breed 2:185–210

Salado-Navarro LR, Sinclair TR, Hinson K (1993) Changes in yield and seed growth traits in soybean cultivars released in the southern USA from 1945 to 1983. Crop Sci 33:1204–1209

Santarém ER, Finer JJ (1999) Transformation of soybean [Glycine max (L.) Merrill] using proliferative embryogenic tissue maintained on semi-solid medium. In vitro Cell Dev Biol-Plant 35:451–455

Santarém ER, Trick HN, Essig JS, Finer JJ (1998) Sonication-assisted *Agrobacterium*-mediated transformation of soybean immature cotyledons: Optimization of transient expression. Plant Cell Rep 17:752–759

Schulze J, Balko C, Zellner B, Koprek T, Hänsch R, Nerlich A, Mendel RR (1995) Biolistic transformation of cucumber using embryogenic suspension cultures: long-term expression of reporter genes. Plant Sci 112:197–206

Senior IJ (1998) Uses of plant gene silencing. Bio/Technol Genet Eng Rev 15:79–119

Shewry PR, Tatham AS, Barro F, Barcelo P, Lazzeri P (1995) Biotechnology of breadmaking: unraveling and manipulating the multi-protein gluten complex. Bio/Technology 13:1185–1190

Shimamoto K, Terada R, Izawa T, Fujimoto H (1989) Fertile transgenic rice plants regenerated from transformed protoplasts. Nature 337:274–276

Shoemaker RC, Amberger LA, Palmer RG, Oglesby L, Ranch JP (1991) Effect of 2,4-dichlorophenoxyacetic acid concentration on somatic embryogenesis and heritable variation in soybean [*Glycine max* (L.) Merr.]. In vitro Cell Dev Biol 27P:84–88

Simmonds DH, Donaldson PA (2000) Genotype screening for proliferative embryogenesis and biolistic transformation of short-season soybean genotypes. Plant Cell Rep 19:485–490

Singh RJ (1993) Plant cytogenetics. CRC Press, Boca Raton, Florida

Singh RJ, Hymowitz T (1988) The genomic relationship between *Glycine max* (L.) Merr. and *G. soja* Sieb. and Zucc. as revealed by pachytene chromosome analysis. Theor Appl Genet 76:705–711

Singh RJ, Hymowitz T (1999) Soybean genetic resources and crop improvement. Genome 42: 605–616

Singh RJ, Klein TM, Mauvais CJ, Knowlton S, Hymowitz T, Kostow CM (1998) Cytological characterization of transgenic soybean. Theor Appl Genet 96:319–324

Sneller CH (1994) Pedigree analysis of elite soybean lines. Crop Sci 34:1515–1522

Somers DA, Torbert KA, Pawlowski WP, Rines HW (1994) Genetic engineering of oat. In: Henry RJ, Ronalds JA (eds) Improvement of cereal quality by genetic engineering. Plenum Press, New York, pp37–46

Songstad DD, Somers DA, Griesbach RJ (1995) Advances in alternative DNA delivery techniques. Plant Cell Tissue Organ Cult 40:1–15

Stam M, Mol JNM, Kooter JM (1997) The silence of genes in transgenic plants. Ann Bot 79:3–12

Stewart CN Jr, Adang MJ, All JN, Boerma HR, Cardineau G, Tucker D, Parrott WA (1996) Genetic transformation, recovery, and characterization of fertile soybean transgenic for a synthetic *Bacillus thuringiensis cryIAc* gene. Plant Physiol 112:121–129

Toriyama K, Arimoto Y, Uchimiya H, Hinata K (1988) Transgenic rice plants after direct gene transfer into protoplasts. Bio/Technology 6:1072–1074

Vasil IK (1994) Molecular improvement of cereals. Plant Mol Biol 25:925–937

Widholm JM (1993) Notice of retraction. Plant Physiol 102:331

Widholm JM (1996) In vitro selection and culture-induced variation in soybean. In: Verma DPS, Shoemaker RC (eds) Soybean genetics, molecular biology and biotechnology. CAB International, Wallingford, UK, pp 107–126

Wright MS, Ward DV, Hinchee MA, Carnes MG, Kaufman RJ (1987) Regeneration of soybean (*Glycine max* L. Merr.) from cultured primary leaf tissue. Plant Cell Rep 6:83–89

Xu X, Li B (1994) Fertile transgenic India rice obtained by electroporation of the seed embryo cells. Plant Cell Rep 13:237–242

Yan B, Srinivasa Reddy MS, Collins GB, Dinkins RD (2000) *Agrobacterium tumefaciens* – mediated transformation of soybean [*Glycine max* (L.) Merrill.] using immature zygotic cotyledon explants. Plant Cell Rep 19:1090–1097

11 Transgenic Barley (*Hordeum vulgare* L.) and Chromosomal Variation

M.-J. Cho, H.W. Choi, P. Bregitzer, S. Zhang, and P.G. Lemaux

11.1 Introduction

Tissue culture and transformation technologies have been developed for most crop species, but several hurdles remain for the efficient application of these methods to crop improvement. One impediment is the occurrence of genetic and stable epigenetic changes during tissue culture and transformation (Lemaux et al. 1999). The objective of transformation should be the introduction of DNA into single, totipotent cells which ultimately give rise to transformed plants that are identical to the parental plant except for the presence of the transgene(s). However, in vitro culture frequently causes genetic and epigenetic changes, including gross cytological changes (e.g., Bayliss 1980; Constantin 1981; Larkin and Scowcroft 1981; Lee and Phillips 1988; Karp 1991; Choi et al. 2000b). These are characterized both as structural rearrangements and numerical variation in the chromosome number. Mechanisms associated with cytological aberrations in vitro include: (1) endoreduplication without cell division in the callus, resulting in polyploid cells, (2) nuclear fragmentation, (3) late replication of heterochromatin causing aneuploidy, (4) chromosomal breakage and (5) structural rearrangements (reviewed by D'Amato 1985; Lee and Phillips 1988; Karp 1991, 1995).

In recent years, barley (*Hordeum vulgare* L.) plants, once considered recalcitrant to genetic engineering approaches, have been successfully transformed using different approaches and explants (reviewed by Lemaux et al. 1999), but transgenic plants had severe reductions in agronomic performance (Bregitzer et al. 1998). Understanding the processes that induce somaclonal variation (SCV) and devising methods to reduce the mutagenicity of the in vitro process involved in transformation will enable the production of superior transgenic plants.

In this chapter, we describe the cytological status of transgenic vs. non-transgenic barley plants and factors affecting cytological aberration during the transformation process, and discuss strategies to minimize SCV and transgene expression instability in transgenic plants – a key goal for the manipulation of crop plants through genetic engineering.

Molecular Methods of Plant Analysis, Vol. 22
Testing for Genetic Manipulation in Plants
Edited by J.F. Jackson, H.F. Linskens, and R.B. Inman
© Springer-Verlag Berlin Heidelberg 2002

11.2 Chromosomal Variation in Nontransgenic Barley Plants

Plants derived from in vitro culture frequently contain heritable genetic and epigenetic changes, termed SCV (Larkin and Scowcroft 1981). In barley these can manifest themselves in many ways, including moderate to severe negative impacts on critical agronomic and quality characteristics (Bregitzer and Poulson 1995; Bregitzer et al. 1995b). The elements of the in vitro environment that induce these changes are poorly understood, but SCV in regenerated nontransgenic plants has been associated with cytogenetic and molecular changes (Phillips et al. 1994). Factors known to induce chromosomal instability in plant tissues during in vitro growth include the particular plant species, genotype, initial ploidy level, explant source, medium composition, growth regulators used, and time in culture (Constantin 1981; Karp 1988).

Numerous reports have described chromosomal variation in cultured tissues and regenerated plants of rice (*Oryza sativa* L.) (Nish and Mitsuoka 1969), wheat (*Triticum aestivum* L.) (Karp and Maddock 1984), maize (*Zea mays* L.) (McCoy and Phillips 1982), oat (*Avena sativa* L.) (McCoy et al. 1982), Italian ryegrass (*Lolium multiflorum* L.) (Jackson and Dale 1988), triticale (X *Triticosecale* Wittmack) (Armstrong et al. 1983; Brettel et al. 1986), *Triticum tauschii* (Winfield et al. 1995), rye (*Secale cereale* L.) (Bebeli et al. 1990) and pearl millet (*Pennisetum americanum* L.) (Swedlund and Vasil 1985). Cytological analyses of cultured barley cells have also shown high frequencies of cytogenetic aberrations, especially polyploidy, but most barley plants derived from in vitro culture in the absence of transformation were diploid (2n = 14) and fertile (Orton 1980; Singh 1986; Karp et al. 1987; Gaponenko et al. 1988; Ziauddin and Kasha 1990; Wang et al. 1992; Hang and Bregitzer 1993; Choi et al. 2000b, 2001a). Choi et al. (2000b) documented a wide range of chromosomal variation in nontransgenic callus cells derived from immature embryos of barley (Table 11.1). The numerical and structural changes in chromosomes occurred in these cells at an early stage; for example, tetraploid cells were observed in scutellar cells as early as 1 day after immature embryos were placed into culture. After 20 weeks in culture, only 42% of cultured cells were diploid; 23% were tetraploid, and 34.9% were aneuploid or had structural variations such as acrocentric, telocentric, and or dicentric chromosomes. Furthermore, nearly 10% of the diploid and tetraploid cells also showed structural variations.

Despite the high frequencies of cytogenetically aberrant callus cells, most barley plants derived from in vitro culture in the absence of transformation were diploid (2n = 2x = 14) and fertile. Karp et al. (1987) reported that 97.6% (41/42) of regenerated barley plants from seven cultivars, including Golden Promise, were diploid (2n = 2x = 14); the one exception was a Golden Promise regenerant which underwent abnormal meiosis. In another study, 99.2% (123/124) of regenerated barley (cv. Moskovsky 3 and 121) plants were diploid (Gaponenko et al. 1988); the remaining plant was tetraploid. Choi et al. (2000b) analyzed chromosome numbers in root tip cells of nontransgenic barley

Table 11.1. Chromosome analysis in nontransgenic barley callus cells from 0 to 84 days post-initiation. (Choi et al. 2000b)

Culturing period (days)[a]	No. cells observed	Percent cells with indicated chromosome number				
		Haploid (7)	Diploid (14)	Tetraploid (28)	Octaploid (56)	Others[b]
0	149	0	100.0	0	0	0
1	115	0	98.3	1.7	0	0
3	109	0	95.4	3.7	0	0.9
7	41	0	97.6	2.4	0	0
14	80	0	93.8	2.5	0	3.7
28	81	0	67.9	8.6	1.2	22.1
84	65	0	42.0	23.1	0	34.9

[a] For callus induction and maintenance, immature embryos were placed and grown on DM medium containing 2.5 mg dicamba/l.
[b] Aneuploid or structural variations.

Table 11.2. Summary of chromosome variation in nontransgenic and transgenic barley plants regenerated from 3-month-old callus. (Choi et al. 2000b)

Nontransgenic/ transgenic	Culture medium[a]	No. plants analyzed	Chromosome number				Percent abnormal ploidy (no. abnormal/total)
			14	26	27	28	
Nontransgenic	DM	6	6	0	0	0	0%
	D	23	22	0	0	1	4.3%
	DC[b]	34	34	0	0	0	0%
	DBC1	10	10	0	0	0	0%
	DBC2	19	19	0	0	0	0%
	Total	92	91	0	0	1	1.1%
Transgenic	DC[b]	138	86	2	4	46	37.7%

[a] Three-month-old calli from each callus-induction medium except for DC medium (callus-induction medium containing 2.5 mg 2,4-D/l and 5.0 μM copper; Cho et al. 1998) were transferred onto FHG regeneration medium (Hunter 1988).
[b] An intermediate culturing step on DBC2 between the callus induction (DC) and regeneration (FHG) step was applied for 30 days.

(cv. Golden Promise) plants regenerated from 3-month-old calli initiated and maintained on each of five different callus-induction media (CIMs), including the media used for the later generation of transgenic callus. Twenty-two plants out of 23 (95.7%) regenerated from D medium (CIM containing 2.5 mg 2,4-D/l; Cho et al. 1998) had diploid numbers, while plants regenerated from all other CIMs were diploid (Table 11.2). Overall, 98.9% (91/92) of nontransgenic, regenerated plants from all media were diploid.

One possible explanation for the preponderance of diploid regenerated plants, given the large number of tetraploid cells in the callus, may be that tetraploid cells are less likely to regenerate than diploid cells. The nontransgenic callus examined in our study derived from tissue that was randomly selected during passage, in contrast to the tissue from which the nontransgenic plants were regenerated, which was selected based on characteristics that correlate with increased regenerability. Consistent with this explanation, Singh (1986) observed that deviations from diploidy were more frequent in nonregenerable (67–68%) vs. regenerable (0–26%) callus cells. Thus, the selection of embryogenic cultures for regenerability would be expected to give a much lower percentage of cells with chromosomal variation.

In the study by Choi et al. (2000b) the choice of culture medium did not contribute to increased variation in chromosome numbers in callus cells of barley. However, a higher percentage of chromosomal aberration was observed in callus cells maintained on CIMs containing auxin in the absence of cytokinin and high levels of copper (D or DM containing 2.5 mg dicamba/l; Wan and Lemaux 1994) than in callus maintained on medium (DBC2, DBC3; Cho et al. 1998) containing 2,4-D in combination with high 6-benzylaminopurine (BAP) and copper (H.W. Choi, P.G. Lemaux and M.-J. Cho, unpubl. data). DBC2 or DBC3 medium is used for induction and maintenance of more organogenic tissues, termed highly regenerative, green tissues (Cho et al. 1998). Analyses of genomic DNA methylation data and field studies of agronomic performance showed lower levels of SCV in barley plants regenerated from highly regenerative, green tissues and cultured shoot meristems than in barley plants regenerated from embryogenic callus tissues initiated and maintained on 2,4-D alone (Zhang et al. 1999b) (see Sect. 11.4.1). Similar trends have been observed in the cytogenetic status of regenerated, nontransgenic oat plants (Choi et al. 2000a). In this study, 44% of plants regenerated from D′ medium containing 2.0 mg 2,4-D/l showed gross cytogenetic abnormalities. Of the plants regenerated from D′BC2 medium containing 2.0 mg 2,4-D/l, 0.1 mg BAP/l, and 5.0 μM copper, 14% were cytologically abnormal. None of the plants regenerated from a D′BC2/DBC3 media regime had cytogenetic abnormalities (DBC3 medium contains 1.0 mg 2,4-D/l, 0.5 mg BAP/l, and 5.0 μM copper). Thus, as the tissues from which plants were regenerated become and are maintained in a more organized and meristematic state, the plants regenerated from them are less likely to have chromosomal variation. This is consistent with a study of maize callus in which greater tissue organization was associated with reductions in SCV (Armstrong and Phillips 1988).

11.3 Chromosomal Variation in Transgenic Barley Plants

In addition to the impact of the tissue culture process alone, the stresses of the transformation process exacerbate SCV. Bregitzer et al. (1998) detected significantly greater reductions in the agronomic performance of progeny of transgenic plants than from regenerated nontransgenic plants. Subsequent cytogenetic studies provided additional evidence that transformation caused SCV over and above that of tissue culture alone for both barley and oat. Chromosomes were studied in cells from callus and root tips of barley plants derived from nontransgenic and transgenic callus of approximately the same age and cultured on the same medium (Choi et al. 2000b, 2001a). Greater variation in the ploidy of transgenic plants was observed compared to nontransgenic plants (Choi et al. 2000b). Of 59 independent transgenic lines, only 32 (54%) had normal diploid complements of $2n = 2x = 14$ (Fig. 11.1A), while 27 (46%) were tetraploid ($2n = 4x = 28$) (Fig. 11.1B) or aneuploid around the tetraploid level (i.e., 26 and 27 chromosomes); no aneuploidy around the diploid number was observed (Table 11.2). One extra, small acrocentric chromosome was detected in an aneuploid T_0 plant (26 chromosomes) and in a tetraploid T_2 plant (28 chromosomes) (Choi et al. 2000b; Fig. 11.1C). Aneuploid plants with 29 and 30 chromosomes were observed in T_1, T_2 and T_3 progeny

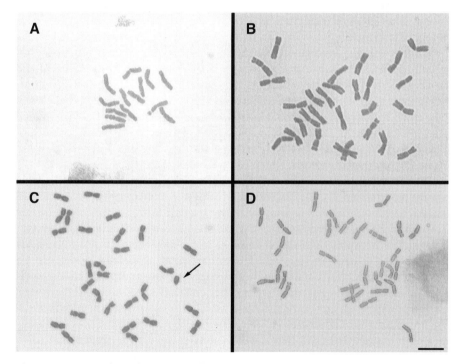

Fig. 11.1A–D. Metaphase chromosomes from root tips of transgenic barley plants with numerical and structural variation. **A** Normal diploid complement of $2n = 2x = 14$. **B** Tetraploid complement of $2n = 4x = 28$. **C** Tetraploid complement of $2n = 4x = 28 + 1$ acrocentric chromosome (*arrow*). **D** Aneuploid complement around the tetraploid level with 30 chromosomes. The transgenic plants were generated via microprojectile bombardment of immature embryos and regenerated after selection with 5 mg bialaphos l^{-1}. *Bar* 10 μm. (Choi et al. 2000b)

from tetraploid plants (Choi et al. 2000b; Fig. 11.1D). Carlson et al. (2001) also reported a relatively high frequency of cytological variation in transgenic barley plants; one (20%) of five transgenic lines was tetraploid ($2n = 4x = 28$). Nontransgenic plants regenerated after in vitro culture alone had a much lower percentage of tetraploids (0–4.3%) (Choi et al. 2000b).

More recently, Choi et al. (2001a) have shown evidence of greater cytological instability in transgenic callus tissues of barley (78% of cells abnormal) than in nontransgenic callus tissues (15% of cells abnormal). Of 22 independently transformed callus lines, only seven had a relatively high percentage (35–76%) of cytologically normal diploid cells ($2n = 2x = 14$). The remaining 15 lines showed a high percentage (92–100%) of cytologically abnormal cells: primarily, tetraploidy ($2n = 4x = 28$), octaploidy ($2n = 8x = 56$), aneuploidy and structural variations. Plants regenerated from transgenic callus tissues with ploidy differences almost always had a chromosomal status comparable to callus tissues from which they derived.

The stresses imposed by the transformation process, above those caused by the in vitro culturing process itself, likely added to or exacerbated the chromosomal instability in tissues from which the transformed plants were regenerated. The DNA introduction process of microprojectile bombardment is potentially stressful. It can involve osmotic treatment, exposure of the cells to a vacuum, cellular damage due to microprojectile impact and potential loss of cell turgor following particle impact. In addition, selection, a part of the process, is used to identify transformed tissue (Jähne et al. 1994; Ritala et al. 1994; Wan and Lemaux 1994; Funatsuki et al. 1995; Hagio et al. 1995; Salmenkallio-Marttila et al. 1995; Koprek et al. 1996; Lemaux et al. 1996; Tingay et al. 1997; Cho et al. 1998, 1999a,b, 2002; Zhang et al. 1999a). During the selection process, transformed tissue must grow in the presence of dead or dying tissue for prolonged periods, which likely causes cellular stress.

The factor(s) in the bombardment-mediated transformation process, principally responsible for the increased cytological aberration in transformed tissues, has recently been examined in the Golden Promise cultivar of barley (Choi et al. 2001a). Only 18 and 21% of 6- and 12-week-old callus cells, respectively, had observable cytological changes in the absence of imposing various transformation stresses (Fig. 11.2). Bombardment alone did not cause an increase in the number of cells with cytological abnormalities. However, this result might be different from that of other cultivars since immature embryos from different cultivars have differing capacities to withstand microprojectile bombardment (Koprek et al. 1996). Damage by microprojectiles to immature embryos of Galena, a two-rowed commercial cultivar recalcitrant to in vitro culture, was found to be significantly greater than damage to immature embryos of Golden Promise, as judged by scanning electron microscopic analysis (Koprek et al. 1999). Koprek et al. (1996) also showed that the callus response of five recalcitrant cultivars was severely reduced by bombardment at pressures around 1,100 psi with the Bio-Rad microparticle gun, although the callus response of Golden Promise was unaffected.

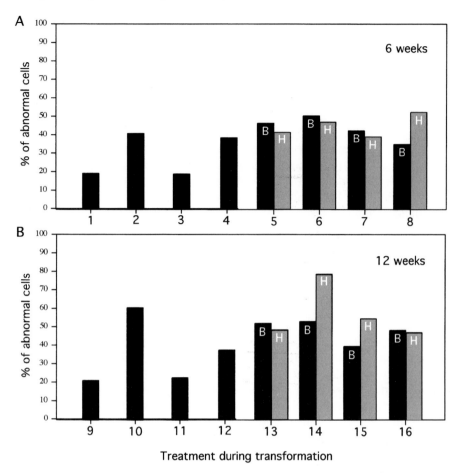

Fig. 11.2A,B. Effect of bombardment, osmoticum and/or selection treatment(s) on cytological aberration in barley callus cells. Callus tissues were manipulated in the following manner and maintained on DC medium for 6 (**A**) and 12 (**B**) weeks after initial callus induction: *1, 9* growth on DC alone, *2, 10* growth on DC + osmotic treatment, *3, 11* growth on DC + bombardment, *4, 12* growth on DC + osmotic treatment + bombardment, *5, 13* growth on DC + selection (2 mg bialaphos/l or 20 mg hygromycin B/l), *6, 14* growth on DC + osmotic treatment + selection (2 mg bialaphos/l or 20 mg hygromycin B/l), *7, 15* growth on DC + bombardment + selection (2 mg bialaphos/l or 20 mg hygromycin B/l), *8, 16* growth on DC + osmotic treatment + bombardment + selection (2 mg bialaphos/l or 20 mg hygromycin B/l). *B* bialaphos selection, *H* hygromycin B selection. (Choi et al. 2001a)

In contrast to the lack of effect of microparticle bombardment, imposing either osmotic or selection treatment on cultured cells caused extensive cytological aberrations, 41 and 40–46% in 6-week-old callus cultures (Fig. 11.2A) and 60 and 48–51% in 12-week-old callus cultures (Fig. 11.2B), respectively. There appears to be no marked difference in the level of cytological aberration caused by either selection or osmotic treatment alone in tissues of either

6 or 12 weeks of age (Choi et al. 2001a; Fig. 11.2). However, if increased levels of selection pressure or longer selection periods are used, selection treatments could cause a higher degree of aberration than the osmotic treatment. For the osmotic treatment, the full extent of that treatment, as applied during actual transformation experiments, was used. For selection a minimum selection scheme was employed so that most callus tissues survived. In the data in Fig. 11.2, callus was cultured for 2 weeks on selection media (2 mg bialaphos/l or 20 mg hygromycin B/l), for 4 weeks without selection and for one more week on selection media. In contrast to this modified selection scheme, a standard selection scheme to obtain transgenic lines involves selection with 5 mg bialaphos/l for about 12 weeks. Thus the extent of the cytological variation in this experiment was probably less than would be seen in tissues undergoing the full selection scheme.

During the first few weeks of the in vitro culture of barley tissue, regenerability typically declines drastically (Jiang et al. 1998) and widespread genomic alterations can be observed, either directly in cultured cells or in regenerated plants. Of 22 independently transformed callus lines, two had high percentages of octaploid cells and other abnormal cells, and were not regenerable (Choi et al. 2001a). However, regenerability was not necessarily related to the degree of visible chromosomal aberrations. For example, two nonregenerable lines had higher percentages of diploid cells, 76 and 73%, than lines that were regenerable (Choi et al. 2001a). In these two cases, other, less visible changes in chromosomal fidelity, such as point mutations, small deletions or insertions or methylation polymorphism, might have affected regenerability. Zhang et al. (1999b) observed that the alterations in methylation patterns positively correlated with the time in culture and negatively correlated with the regeneration potential of the cultures.

Studies of other species indicate that cytological aberrations are common in transgenic cereal and grass plants. Frequent changes in ploidy level were observed in transgenic orchardgrass plants (Cho et al. 2001). Plants from only three lines (30%) had a normal tetraploid number of chromosomes ($2n = 4x = 28$), while plants from seven lines (70%) were octaploid ($2n = 8x = 56$). None of the 15 tissue culture-derived or 20 seed-derived nontransgenic plants had abnormal ploidy. Another dramatic example of the effect of the transformation process on gross chromosomal instability was observed in oat, a hexaploid ($2n = 6x = 42$) species. Sixty percent of transgenic plants had cytogenetic aberrations compared to 0–14% in nontransgenic oat plants cultured on the same medium for the same length of time (Choi et al. 2000a). The most common cytogenetic aberration was aneuploidy ($2n = 6x = 40, 41$), followed by deletion of chromosomal segments; no change in ploidy level was observed. Similar results were observed in wheat with transgenic hexaploid plants ($2n = 6x = 42$), i.e., 36% abnormality vs. 10% with nontransgenic plants (Choi et al. 2001c). A last example is the frequent aneuploidy (50%) observed in transgenic plants from a hexaploid tall fescue species (*Festuca arundinacea* Schreb.) ($2n = 6x = 42$) compared with nontransgenic plants (17%); no change in ploidy level was observed (Choi et al. 2001c). The conclusion from all of these studies is that the nature of the chromosomal aberration, e.g., strict ploidy change or aneuploidy, appears to be dependent upon the particular plant species and its fundamental genomic state. The changes observed in chromosomal number and integrity of chromosomes only permit quantitation of gross changes in chromosomal fidelity. It is likely that other, less visible changes in chromosomal fidelity occur, e.g., point mutations, short deletions or additions and methylation polymorphism (Phillips et al. 1994; Zhang et al. 1999b). These changes likely impair the ability of the transgenic plants to grow, perform and reproduce in a manner identical to the nontransgenic parental plants.

11.4 Fidelity and Quality of Transgenic Barley Plants

SCV is an inherent characteristic of in vitro culture systems and will likely influence the application of genetic engineering to plant improvement. The impact might be observed in at least four ways: (1) genotypic and temporal restrictions on regeneration of fertile, green plants; (2) alteration of the expression of carefully selected, commercially valuable characteristics; (3) gross genomic alterations, including ploidy changes; and (4) instability in transgene expression and inheritance.

11.4.1 Comparative Analysis of Genomic Stability in Plants Derived from Tissues Generated Using Different in Vitro Proliferation Processes

Genome stability of plants derived from in vitro culture is of critical importance for the direct application of transformation technologies to plant improvement. Analysis of previously published studies of barley plants derived from embryogenic callus shows considerable variation not only in terms of karyotype and chromosome stability (see Sect. 11.3) but also in agronomic performance. Ullrich et al. (1991) observed genotype-dependent morphological variation among plants regenerated from 18 cultivars or breeding lines of barley. In another field study, protoplast-derived barley plants from the cultivars Igri and Dissa showed significant negative changes in height, heading date, fertility, spike length, and spikelet density (Kihara et al. 1997) compared to nontissue culture-derived plants. In studies of agronomic performance (Bregitzer and Poulson 1995) and malting quality (Bregitzer et al. 1995b), significant, undesirable, and genotype-dependent changes were detected in plants from the majority of the 30 analyzed tissue culture-derived lines derived from six North American barley cultivars, including Golden Promise and Morex. Collectively, the effects of SCV in transgenic plants are likely to have a negative impact on attempts to improve commercial germplasm using genetic engineering technologies.

A study was initiated to develop molecular methods that could be used to predict the severity of the effects of in vitro culture on plant quality and performance. The methylation state of the genome in uncultured control plants was compared to those derived from different in vitro culture methods (Zhang et al. 1999b) that were used previously to transform barley. These include the standard embryogenic callus induction from immature embryos (Wan and Lemaux 1994), highly regenerative green tissue from a modified culturing method (Cho et al. 1998) and in vitro shoot meristem cultures (Zhang et al. 1999a). Single plant-derived seed of two cultivars, Golden Promise and Morex, was used to initiate the three different types of in vitro cultures in order to avoid potential complications caused by heterogeneity within the cultivar. Two time-points were chosen for plant regeneration, 1 and 3 months for the embryogenic and highly regenerative cultures and 3 and 6 months for the shoot meristem cultures. Genomic DNA from regenerated plants was isolated and analyzed using methylation-sensitive enzymes and random genomic probes (Zhang et al. 1999b).

In Golden Promise plants, the patterns of DNA hybridization in the majority of the plants derived from the standard embryogenic callus route were different from those in control plants. Also the frequency of methylation changes in plants from embryogenic cultures was more marked in plants regenerated at 3 months than at 1 month. In contrast, analysis of the results from plants derived from shoot meristem cultures, even at 6 months, showed that hybridization patterns were relatively stable compared to those deriving from the standard embryogenic culturing procedure. In most Golden Promise plants, the hybridization patterns of DNA from plants obtained from the shoot meristem cultures were identical to those from single plant-derived seed. DNA from plants deriving from highly regenerative cultures showed intermediate levels of methylation polymorphism between those from shoot meristem culture and the standard embryogenic approach. The lower frequency of methylation polymorphism in plants from highly regenerative cultures than in those from the standard embryogenic method might be due to the fact that in highly regenerative cultures early conversion of tissue from an embryogenic mode to a more meristematic mode of growth occurs. It is possible that the meristematic growth mode results in more genomic stability during in vitro culture.

The reduction in methylation polymorphism might be related to the mode of in vitro growth, i.e., tissues undergoing less dedifferentiation might experience less stress, which manifests itself as a lower rate of methylation change. It has been proposed that methylation changes are a major factor in SCV (Phillips et al. 1994). The lower rate of methylation change in plants deriving from the meristematic cultures is likely to lead to reduced mutations, and therefore to lower impacts on field performance (Zhang et al. 1999b). The results of agronomic evaluations of the resulting Golden Promise plants were consistent with the methylation data, in that agronomic performance showed some correlation with the degree of methylation variation (Bregitzer et al. 2002).

The conclusion from the preliminary analyses of methylation polymorphism and agronomic performance data is that the use of shoot meristem cultures or the earlier conversion of embryogenic cultures to a meristematic growth mode reduces methylation polymorphism and may improve field performance of in-vitro-derived plants. The use of more meristematic or organogenic cultures represents a departure from many published procedures for in vitro culturing and transformation since much effort has been expended on developing and characterizing embryogenic cultures of cereals as useful targets for transformation efforts (e.g., Vasil 1987). These data on methylation stability and SCV suggest that the focus justifiably might now shift to cultures that are propagated in a meristematic state. While SCV is likely reduced in plants derived from meristematic cultures, it has also been observed that a substantial increase in SCV occurs in response to transformation attempts, documented to date in field tests of transformed rice and barley plants.

11.4.2 Somaclonal Variation and Field Performance of Transgenic Plants Derived from Embryogenic Callus

The original transgenic lines of Wan and Lemaux (1994), carrying *uidA, bar* and a gene encoding the coat protein of barley yellow-dwarf virus, were characterized by abnormally slow growth and development during the initial generation and seed increases in the greenhouse. Poor growth of this kind had not been previously observed in plants of any genotype that had derived from in vitro culturing alone (P. Bregitzer, unpubl. data) and was early evidence of a significant problem with the quality of transgenic barley plants. Phenotypic variation in transgenic plants is also related to changes in ploidy level (Choi et al. 2000b, 2001a). Tetraploid transgenic plants were different from diploid transgenic plants in growth rate and morphology at both the vegetative and reproductive stages (Fig. 11.3A) with later heading dates, broader leaves, thicker roots and longer seeds in tetraploid plants. Similar phenotypic abnormalities, such as larger plant organs in polyploid than in diploid plants, were reported in regenerated nontransgenic plants of alfalfa and potato (reviewed by Lee and Phillips 1988). A low frequency of an abnormal grass-like phenotype with very short height and sterility was observed mostly in aneuploid T_1 barley plants around the tetraploid level (Choi et al. 2001a) and sometimes in diploid T_1 plants (Fig. 11.3B).

Further examination in field plots of plants from the first barley lines (Wan and Lemaux 1994) derived from transgenic plants provided some disturbing data (Bregitzer et al. 1998). Results with null (nontransgenic) segregants were studied in order to determine that the observed alterations in agronomic performance were attributable to SCV, and not confounded by potential effects of the insertion or expression of the transgenes. Morphological abnormalities were more common in null segregants derived from transgenic Golden Promise callus lines than in those from tissue-culture-derived, nontransgenic

Fig. 11.3A,B. Phenotypic abnormalities in transgenic barley plants. **A** Difference in growth rate between diploid (*left*) and tetraploid (*right*) plants at the vegetative stage. Tetraploid plants were delayed in seed maturation and had thicker, wider leaves (Choi et al. 2000b). **B** Abnormal grass-like phenotype with very short height and sterility (*middle* and *right*) was observed in some diploid T_1 plants

Golden Promise plants (Bregitzer and Poulson 1995). Agronomic data also indicated that transformation imposed additional levels of SCV. Heading date was delayed, plants were shorter and lower yielding, and the seed weight was reduced in the transgenic lines relative to those derived from the tissue culture process alone. For example, grain yield of transgenic Golden Promise plants was approximately 50% of the uncultured Golden Promise control in field trials, whereas the grain yield of nontransgenic, tissue-culture-derived Golden Promise lines averaged over 90% of the uncultured control. These observations are consistent with a similar study of rice, in which control plants and protoplast-derived nontransgenic rice plants had similar agronomic characteristics (Schuh et al. 1993). Protoplast-derived transgenic plants, however, were markedly inferior, e.g., yield was only 10% of the uncultured control.

Examination of the agronomic characteristics of transgenic-derived lines showed that the T_2 and T_4 generations were indistinguishable in their heading date, height, yield or seed weight (Bregitzer et al. 1998), indicating that the observed SCV was also heritable in self-pollinated progeny and no amelioration of the SCV with generation advance was observed. Several studies of regenerated maize plants also indicated that quantitative and qualitative SCV was heritable (reviewed by Kaeppler and Phillips 1993a,b).

11.4.3 Stability of Transgenes and Transgene Expression

Stable physical transmission and expression of transgenes are critical for efficient application of transformation technologies to plant breeding.

However, the use of current transformation methods, mediated by either T-DNA or direct gene transfer, often leads to plants in which the transgene itself or its expression is unstable (reviewed by Finnegan and McElroy 1994). Evidence of both physical loss of the transgene and instability of transgene expression has been well-documented in barley. In the original report of stable transformation, 69% (24/35) of the T_0 plants, representing 21 independent lines, did not show Mendelian inheritance of expression in T_1 progeny (Wan and Lemaux 1994). In addition, evidence of loss of the actual transgenes in T_1 or later generation progeny has also been documented in a subset of the events (Cho et al. 2002; T. Koprek , R. Williams-Carrier, R. Fessenden, P.G. Lemaux and P. Bregitzer, unpubl. data). The mechanisms involved in the physical loss (partial or complete) or amplification of transgenes in later generation plants are not well characterized.

Mechanisms to explain transgene expression variability have been put forward and these involve, to a large extent, homology-dependent or repeat-induced gene silencing (for review, see Matzke and Matzke 1995). These mechanisms can be post-transcriptional (co-suppression) or transcriptional, often associated with DNA methylation at repeat sequences. These explanations appear not to be adequate to explain all cases of transgene expression variability, such as those seen among clonal progeny of single T_0 plants or the differing rates at which transgene silencing occurs among clonal progeny. These explanations are also not consistent with the observed lack of direct correlation between the severity of silencing and the extent of sequence similarity (Conner et al. 1997), with the lack of linear correlation between the number of copies integrated and the severity of transgene silencing (e.g., Wan and Lemaux 1994) or the evidence of endogenous gene silencing in null-segregant progeny of transgenic events (Conner et al. 1997).

Most current cereal transformation methods use in-vitro-proliferated target cells which incur methylation pattern changes, chromosomal rearrangements/ deletions, ploidy changes and genetic mutations (Hang and Bregitzer 1993; Phillips et al. 1994; Zhang et al. 1999b; Choi et al. 2000a,b; Cho et al. 1999a, 2002; Carlson et al. 2001). These changes could lead to a breakdown in normal cellular control mechanisms which can further exacerbate transgene instability. That genomic instability of in vitro cultured plant cells might affect the stability of transgenes and their expression is exemplified by a study of the integration-site structure in transformants created by direct gene transfer to rice protoplasts (Takano et al. 1997). Transgenes were found in regions with inverted structures and large genomic duplications; very limited sequence homology between plasmid and target DNA implicated a role for illegitimate recombination in the process. Since only three junctions were studied and each contained extensive rearrangements, it was suggested that the process of in vitro culturing of the target protoplasts was responsible for predisposing the cells to illegitimate recombination and genomic rearrangements. In addition, in another study on gene targeting using T-DNA, the target loci were deleted in up to 20% of the calli deriving from tobacco leaf protoplasts (Risseeuw et al. 1997). A possible explanation is that the T-DNAs become unstable during in vitro propagation and undergo deletions. In stressed cells, such as callus, genomic instability occurs at particular loci and these sites become favored targets for T-DNA integration by illegitimate recombination methods. Analysis of the results from these two studies indicates that in vitro culturing leads to genomic instability and suggests that using cells with relatively stable genomes might improve transgene stability and expression.

In barley, stability of the transgene itself and its expression have been studied in transgenic plants. These plants derived from independently transformed lines from the standard embryogenic culture approach (Wan and Lemaux 1994) and were analyzed using Southern blot, PCR and in vivo biochemical assays to determine the physical presence or expression of the transgenes (P. Bregitzer, R. Williams-Carrier, P.G. Lemaux, unpubl. data). Plants from some lines were characterized as showing Mendelian inheritance of the transgene and its expression through four generations, but some lines showed significantly fewer transgene-expressing segregants than expected. In the unstably expressing transgenic lines, as well as in a number of earlier-generation plants from other transgenic lines, it is likely that the variation in expression stability among the different transgenic events is related at least in part to the site of transgene integration. The influence of the site could be mediated by the fact that different genomic methylation patterns are present in embryogenic vs. mature plant tissue. This could lead to a situation in which a transgene integrated into a site not methylated in the embryo, but is subsequently methylated in a mature tissue. Another mediating factor could be the fact that demethylation can occur during in vitro culture (Phillips et al. 1994). This could lead to a situation in which a transgene integrates into a site that has been demethylated in the T_0 plant, for example; however, the site becomes remethylated in subsequent generations. The propensity for and speed with which any one site will undergo demethylation and remethylation could be different in different transgenic lines, leading to variation in the propensity of an integrated transgene to undergo silencing due to methylation. In addition, the rate at which remethylation occurs would also likely vary among clonal progeny, leading to different rates of observed transgene silencing among plants from the same line in the same generation. This could explain the observations of Zhang et al. (1996).

Analyses must be carried out to look at the characteristics of transgene integration sites in lines which are stable over multiple generations vs. those showing physical or expression instability. A correlation between the degree of site methylation and the stability or instability of the transgene and the region surrounding the transgene as well as its expression could be useful in predicting transgene stability. In barley, these analyses could take advantage of the fact that cereal genomes are known to contain large percentages of repeat DNA (80% in barley), probably generated through non-random amplification of DNA segments that are hypermethylated (Moore 1995). This highly repeated, hypermethylated DNA is interspersed with undermethylated sequences that contain genes. Certain of these elements in the barley genome are amplifiable, prone to methylation and localized to distinct restriction-fragment size classes (Moore et al. 1991). It has also been noted that recombination in large cereal genomes is predominantly confined to regions distal to the centromere (Moore et al. 1991). Stability of the transgene and its expression could be compared to the proximity of the transgene to repeat DNA and to the centromere. A correlation between the stability of the transgene and the characteristics of its integration site could enable the development of diagnostic tools that are useful in predicting the stability of the transgene and its expression in early-stage transformants.

The segregation ratio of transgene expression is also related to the ploidy level of transgenic barley plants. Theoretically, diploid transgenic plants have a 3:1 transgene expression segregation ratio for transgenes with dominant gene action. In contrast, two different segregation patterns of transgene expression are possible in tetraploid transgenic barley plants. If cells were already tetraploid at the time of DNA integration, it would be expected that transgene segregation would be 3:1; if cells were diploid and became tetraploid after DNA integration, then a 35:1 segregation ratio would be expected. The timing of becoming tetraploid or aneuploid in transgenic barley is influenced by the fact that DNA introduced by bombardment can remain in the cultured cells for at least 2 weeks post-bombardment (unpublished data). This can lead to an increasing probability of the transgene integrating into cells that have already become tetraploid, especially since the stresses of the transformation process exacerbate ploidy instability.

This hypothesis was confirmed using fluorescence in situ hybridization (FISH) on the metaphase chromosomes of transgenic barley transformed with *uidA* and *sgfp*(S65T) genes (Choi et al. 2002). Of the 14 independent transgenic lines, six were *uidA*-containing and eight contained *sgfp*(S65T) (3 diploid and 11 tetraploid lines). Only a single integration site was detected on one of homologous chromosomes in diploid ($2n = 2x = 14$) T_0 plants of events with a 3:1 segregation ratio; homozygous plants obtained in T_1 or later generations of these events had doublet signals on a pair of homologous chromosomes. Our FISH analyses showed that tetraploid ($2n = 2x = 28$) T_0 plants with a 3:1 segregation ratio also had only a single integration site on a one of homologous chromosomes; this indicates that cells were already tetraploid at the time of DNA integration. In contrast, tetraploid T_0 plants with a 35:1 segregation ratio in the T_1 generation had doublet signals on a pair of homologous chromosomes, indicating that cells were diploid at the time of integration and later became tetraploid. One tetraploid T_0 line with a segregation ratio consistent with being a homozygote (45:0) had doublet signals at two loci on separate chromosomes.

Diploid transgenic plants were found to have a higher physical loss and inactivation of expression of the transgenes *sgfp*(S65T) and *bar*, driven by the rice actin and maize ubiquitin promoters, respectively, than did tetraploid transgenic plants (Cho et al. 2002). Less than five out of the ten diploid lines tested expressed either transgene. All twelve tetraploid lines tested showed stable transgene expression, although transgene silencing was also observed in some progeny of two tetraploid lines. The lower frequency of transgene silencing in tetraploid plants relative to diploid plants might be due to the higher buffering capacity for chromosome damage in tetraploid transgenic lines.

Transgene expression stability is also dependent upon the type of promoter. Expression of GFP driven by the D-hordein promoter was much more stable in its inheritance pattern than that driven by the rice actin promoter or *bar* driven by the maize ubiquitin promoter (Cho et al. 2002). GFP expression driven by the D-hordein promoter was stably transmitted to T_2 progeny of all seven fertile *sgfp*(S65T)-positive independent lines tested. In contrast to hordein-driven expression, at least five lines out of 14 independent lines lost GFP expression driven by the rice actin promoter. Similarly, expression of *bar* driven by the maize ubiquitin promoter was lost in T_1 progeny; only 21 out of 26 independent lines were Basta-resistant. These results are also similar to our previous results (Cho et al. 1999a), which showed that the barley B_1- and D-hordein promoters provide more stable expression of GUS during generation advance than expression from the maize ubiqutin-driven *bar* gene. The patterns of the tissue-specific gene expression and segregation ratios of GUS- or GFP-expressing lines driven by B_1- or D-hordein promoter were inherited in plants of T_4–T_8 progeny of each transgenic line tested (Choi et al. 2001b).

The type of transgene also appears to have an effect on transgene stability and its inheritance. For example, there was higher transgene expression instability in GFP-expressing diploid lines driven by actin than in ubiquitin-driven PAT-expressing diploid lines (Cho et al. 2002). All six (100%) diploid GFP-expressing lines lost GFP expression at the T_1 generation, either due to transgene loss or transgene inactivation. This could be explained by mild phytotoxicity of GFP (Haseloff et al. 1997), which can cause a higher loss of physical transmission of the transgene. Five (50%) out of the ten diploid lines lost Basta-resistance in their T_1 progeny (Cho et al. 2002). In some earlier reports in oat, loss or low rates of physical transmission of the transgene(s) to progeny were also observed (Somers et al. 1994; Pawlowski and Somers 1996; Pawlowski et al. 1998).

11.5 Conclusions and Future Perspectives

Plants from in vitro culture can exhibit SCV, two characteristics of which are structural rearrangements and numerical variation in the chromosomes. Reduced SCV in barley is associated with culturing highly differentiated, meristematic tissues; the use of more organogenic cultures can reduce methylation polymorphism and improve field performance of in vitro-derived plants. The extent of chromosomal aberrations in transgenic callus and plants is exacerbated due to the additional stresses that occur during transformation, e.g., selection and osmotic treatment. We have recently observed a much higher frequency (46%) of chromosomal abnormality in transgenic barley plants than in nontransgenic plants (0–4.3%) regenerated from tissues of comparable age cultured on identical media.

Therefore, reducing the stresses associated with the transformation process and using more organogenic cultures as transformation targets appear to be desirable in order to generate transgenic plants that are more genetically and agronomically identical to the parents, a key goal for the manipulation of crop plants through genetic engineering. If this goal is not achieved, additional time-consuming back-crossing to parental germplasm has to be undertaken in order to eliminate undesirable induced mutations. In some cases induced mutations are closely linked to the transgene(s) and cannot be eliminated. Even if back-crossing can be used, the process usually only slows down the generation of transgenic plants that are agronomically identical to the parental germplasm. Transformation methods that minimize the impact of osmotic, selection and other stresses in combination with the use of more organogenic tissues should lead to the regeneration of plants with minimal evidence of SCV.

References

Armstrong CL, Phillips RL (1988) Genetic and cytogenetic variation in plants regenerated from organogenic and friable embryogenic tissue culture of maize. Crop Sci 28:363–369

Armstrong KC, Nakamura C, Keller WA (1983) Karyotype instability in tissue culture regenerants of triticale (X *Triticosecale* Wittmack) cv. 'Welsh' from 6-month old callus cultures. Z Pflanzenzuecht 91:233–245

Bayliss MW (1980) Chromosomal variation in plant tissues in culture. In:Vasil IK (ed) Int Rev Cytol (Suppl). Academic Press, New York, pp 113–144

Bebeli PJ, Karp A, Kaltsikes PJ (1990) Somaclonal variation from cultured immature embryos of sister lines of rye differing in heterochromatic content. Genome 33:177–183

Bregitzer P, Poulson M (1995) Agronomic performance of barley lines derived from tissue culture. Crop Sci 35:1144–1148

Bregitzer P, Campbell RD, Wu Y (1995a) Plant regeneration from barley callus: effects of 2,4-dichlorophenoxyaceticacid and phenylacetic acid. Plant Cell Tiss Org Cult 43:229–235

Bregitzer P, Poulson M, Jones BL (1995b) Malting quality of barley lines derived from tissue culture. Cereal Chem 72:433–435

Bregitzer P, Halbert SE, Lemaux PG (1998) Somaclonal variation in the progeny of transgenic barley. Theor Appl Genet 96:421–425

Bregitzer P, Zhang S, Cho M-J, Lemaux PG (2002) Reduced somaclonal variation in barley is associated with culturing highly differentiated, meristematic tissues. Crop Sci (in press)

Brettel RIS, Pallotta MA, Gustafson JP, Appels R (1986) Variation at the *Nor* loci in triticale derived from tissue culture. Theor Appl Genet 71:637–643

Carlson AR, Letarte J, Chen J, Kasha KJ (2001) Visual screening of microspore-derived transgenic barley (*Hordeum vulgare* L.) with green-fluorescent protein. Plant Cell Rep 20:331–337

Cho M-J, Jiang W, Lemaux PG (1998) Transformation of recalcitrant barley cultivars through improvement in regenerability and decreased albinism. Plant Sci 138:229–244

Cho M-J, Choi HW, Buchanan BB, Lemaux PG (1999a) Inheritance of tissue-specific expression of hordein promoter-*uidA* fusions in transgenic barley plants. Theor Appl Genet 98:1253–1262

Cho M-J, Wong JH, Marx C, Jiang W, Lemaux PG, Buchanan BB (1999b) Overexpression of thiore-doxin *h* leads to enhanced activity of starch debranching enzyme (pullulanase) in barley grain. Proc Natl Acad Sci USA 96:14641–14646

Cho M-J, Choi HW, Lemaux PG (2001) Transgenic T_0 orchardgrass (*Dactylis glomerata* L.) plants produced from highly regenerative tissues derived from mature seeds. Plant Cell Rep 20: 318–324

Cho M-J, Choi HW, Jiang W, Ha C, Lemaux PG (2002) Endosperm-specific expression of green fluorescent protein driven by the hordein promoter is stably inherited in transgenic barley (*Hordeum vulgare* L.) plants. Physiol Plant (in press)

Choi HW, Lemaux PG, Cho M-J (2000a) High frequency of cytogenetic aberration in transgenic oat (*Avena sativa* L.) plants. Plant Sci 156:85–94

Choi HW, Lemaux PG, Cho M-J (2000b) Increased chromosomal variation in transgenic versus nontransgenic barley (*Hordeum vulgare* L.) plants. Crop Sci 40:524–533

Choi HW, Lemaux PG, Cho M-J (2001a) Selection and osmotic treatment exacerbate cytological aberrations in transformed barley (*Hordeum vulgare*). J Plant Physiol 158:935–943

Choi HW, Lemaux PG, Cho M-J (2001b) Long-term stability of transgene expression driven by barley endosperm-specific hordein promoters in transgenic barley (*Hordeum vulgare* L.). Cong In Vitro Biol:1110

Choi HW, Lemaux PG, Cho M-J (2001c) Transformation process exacerbates cytological variation in transgenic grass and cereal plants. Cong In Vitro Biol:1108

Choi HW, Lemaux PG, Cho M-J (2002) Use of fluorescence in situ hybridization for gross mapping of transgenes and screening for homozygous plants in transgenic barley (*Hordeum vulgare* L.). Theor Appl Genet (in press)

Conner JA, Stein T, Tantikanjana C, Kandasamy MK, Nasrallah JB, Nasrallah ME (1997) Transgene-induced silencing of *S*-locus genes and related genes of *Brassica*. Plant J 121:809–823

Constantin MJ (1981) Chromosome instability in cell and tissue cultures and regenerated plants. Environ Exp Bot 21:359–368

D'Amato F (1985) Cytogenetics of plant cell and tissue cultures and their regenerates. Critical Rev in Plant Sci. CRC, Boca Raton, pp 73–112

Finnegan J, McElroy D (1994) Transgene inactivation: plants fight back! Bio/Technology 12: 883–888

Funatsuki H, Kuroda H, Kihara M, Lazzeri PA, Müller E, Lörz H, Kishinami I (1995) Fertile transgenic barley generated by direct DNA transfer to protoplasts. Theor Appl Genet 91:707–712

Gaponenko AK, Petrova TF, Iskakov AR, Sozinov AA (1988) Cytogenetics of in vitro cultured somatic cells and regenerated plants of barley (*Hordeum vulgare* L.). Theor Appl Genet 75: 905–911

Hagio T, Hirabayashi T, Machii H, Tomotsune H (1995) Production of fertile transgenic barley (*Hordeum vulgare* L.) plant using the hygromycin-resistance marker. Plant Cell Rep 14: 329–334

Hang A, Bregitzer P (1993) Chromosomal variations in immature embryo-derived calli from six barley cultivars. J Hered 84:105–108

Haseloff J, Siemering KR, Prahser DC, Hodge S (1997) Removal of a cryptic intron and subcel-lular localization of green fluorescent protein are required to mark transgenic *Arabidopsis* plants brightly. Proc Natl Acad Sci USA 94:2122–2127

Hunter CP (1988) Plant regeneration from microspores of barley, *Hordeum vulgare*. PhD Thesis, Wye College, University of London

Jackson JA, Dale PJ (1988) Callus induction, plant regeneration and an assessment of cytological variation in regenerated plants of *Lolium multiflorum* L. J Plant Physiol 132:351–355

Jähne A, Becker D, Brettschneider R, Lörz H (1994) Regeneration of transgenic, microspore-derived, fertile barley. Theor Appl Genet 89:525–533

Jiang W, Cho M-J, Lemaux PG (1998) Improved callus quality and prolonged regenerability in model and recalcitrant barley (*Hordeum vulgare* L.) cultivars. Plant Biotechnol 15:63–69

Kaeppler SM, Phillips RL (1993a) DNA methylation and tissue culture-induced variation in plants. In Vitro Cell Dev Biol 29:125–130

Kaeppler SM, Phillips RL (1993b) Tissue culture-induced DNA methylation variation in maize. Proc Natl Acad Sci USA 90:8773–8776

Karp A (1988) Origins and causes of chromosome instability in plant tissue culture and regeneration. In: Brandham PE (ed) Kew Chromosome Conference III. Her Majesty's Stationery Office, London, pp 185–192

Karp A (1991) On the current understanding of somaclonal variation. Oxford Surv Plant Mol Cell Biol 7:1–59

Karp A (1995) Somaclonal variation as a tool for crop improvement. Euphytica 85:295–302

Karp A, Maddock SE (1984) Chromosome variation in wheat plants regenerated from cultured immature embryos. Theor Appl Genet 67:249–255

Karp A, Steele SH, Parmar S, Jones MGK, Shewry PR (1987) Relative stability among barley plants regenerated from cultured immature embryos. Genome 29:405–412

Kihara M, Okada Y, Kuroda H, Saeki K, Ito K, Yoshigi N (1997) Generation of fertile transgenic barley synthesizing thermostable β-amylase. 26th European Brewing Convention Congress, Maastricht, Netherlands, Oct 20–21, 1997. J Inst Brewing 103:153

Koprek T, Haensch R, Nerlich A, Mendel RR, Schultz J (1996) Fertile transgenic barley of different cultivars obtained by adjustment of bombardment conditions to tissue response. Plant Sci 119:79–91

Koprek T, Cho M-J, Choi HW, Kim HK, Zhang S, Bregitzer P, Lemaux PG (1999) Clearing the hurdles of small grain transformation: a molecular and cytological approach. Proc 9th Australian Barley Technical Symposium, Melbourne, Australia, pp 2.3.1–2.3.8

Larkin PJ, Scowcroft WR (1981) Somaclonal variation – a novel source of variability from cell cultures for plant improvement. Theor Appl Genet 60:197–214

Lee M, Phillips RL (1988) The chromosomal basis of somaclonal variation. Annu Rev Plant Physiol Plant Mol Biol 39:413–437

Lemaux PG, Cho M-J, Louwerse J, Williams R, Wan Y (1996) Bombardment-mediated transformation methods for barley. Bio-Rad US/EG Bulletin 2007:1–6

Lemaux PG, Cho M-J, Zhang S, Bregitzer P (1999) Transgenic cereals: *Hordeum vulgare* L. (barley). In: Vasil IK (ed) Molecular improvement of cereal crops. Kluwer, Dordrecht, pp 255–316

Matzke MA, Matzke AJM (1995) How and why do plants inactivate homologous (trans) genes? Plant Physiol 107:679–685

McCoy TJ, Phillips RL (1982) Chromosome stability in maize (*Zea mays*) tissue cultures and sectoring in some regenerated plants. Can J Genet Cytol 24:559–565

McCoy TJ, Phillips RL, Rines HW (1982) Cytogenetic analysis of plants regenerated from oat (*Avena sativa*) tissue cultures; high frequency of partial chromosome loss. Can J Genet Cytol 24:27–50

Moore G (1995) Cereal genome evolution: pastoral pursuits with "Lego" genomes. Curr Opin Genet Dev 5:717–724

Moore G, Cheung W, Schwarzacher T, Flavell R (1991) BIS 1, a major component of the cereal genome and a tool for studying genomic organization. Genomics 10:469–476

Nish T, Mitsuoka S (1969) Occurrence of various ploidy plants from anther and ovary culture of rice plant. Jpn J Genet 44:341–346

Orton TJ (1980) Chromosome variability in tissue cultures and regenerated plants of *Hordeum*. Theor Appl Genet 56:101–112

Pawlowski WP, Somers DA (1996) Transgene inheritance in plants genetically engineered by microprojectile bombardment. Mol Biotechnol 6:17–30

Pawlowski WP, Torbert KA, Reins HW, Somers DA (1998) Irregular patterns of trangene silencing in allohexaploid oat. Plant Mol Biol 38:597–607

Phillips RL, Kaeppler SM, Olhoft P (1994) Genetic instability of plant tissue cultures: breakdown of normal controls. Proc Natl Acad Sci USA 91:5222–5226

Risseeuw E, Franke-Van Dijk MEI, Hooykaas PJJ (1997) Gene targeting and instability of *Agrobacterium* T-DNA loci in the plant genome. Plant J 11:717–728

Ritala A, Aspegren K, Kurtén U, Salmenkallio-Marttila M, Mannonen L, Hannus R, Kauppinen V, Teeri TH, Enari TM (1994) Fertile transgenic barley by particle bombardment of immature embryos. Plant Mol Biol 24:317–325

Salmenkallio-Marttila M, Aspegren K, Åkerman S, Kurtén U, Mannonen L, Ritala A, Teeri TH, Kauppinen V (1995) Transgenic barley (*Hordeum vulgare* L.) by electroporation of protoplasts. Plant Cell Rep 15:301–304

Schuh W, Nelson MR, Bigelow DM, Orum TV, Orth CE, Lynch PT, Eyles PS, Blackhall NW, Jones J, Cocking EC, Davey MR (1993) The phenotypic characterisation of R-2 generation transgenic rice plants under field conditions. Plant Sci 89:69–79

Singh RJ (1986) Chromosomal variation in immature embryo derived calluses of barley (*Hordeum vulgare* L.). Theor Appl Genet 72:710–716

Swedlund B, Vasil IK (1985) Cytogenetic characterization of embryogenic callus and regenerated plants of *Pennisetum americanum* (L.) K. Schum. Theor Appl Genet 69:575–581

Somers DA, Torbert KA, Pawlowski WP, Reins HW (1994) Genetic engineering of oat plants. In: Improvement of cereal quality by genetic engineering. Plenum, New York, pp 37–46

Takano M, Egawa H, Ikeda J, Wakasa K (1997) The structures of integration sites in transgenic rice. Plant J 11:353–361

Tingay S, McElroy D, Kalla R, Fieg S, Wang M, Thornton S, Brettell R (1997) *Agrobacterium tumefaciens*-mediated barley transformation. Plant J 11:1369–1376

Ullrich SE, Edmiston JM, Kleinhofs A, Kudrna DA, Maatougui MEH (1991) Evaluation of somaclonal variation in barley. Cereal Res Comm 19:245–260

Vasil IK (1987) Developing cell and tissue culture systems for the improvement of cereal and grass crops. J Plant Physiol 128:193–218

Wan Y, Lemaux PG (1994) Generation of large numbers of independently tranformed fertile barley plants. Plant Physiol 104:37–48

Wang XH, Lazzeri PA, Lörz H (1992) Chromosomal variation in dividing protoplasts derived from cell suspensions of barley (*Hordeum vulgare* L.). Theor Appl Genet 85:181–185

Winfield MO, Karp A, Lazzeri PA, Davey MR (1995) Chromosome 5 D instability in cell lines of *Triticum tauschii* and morphological variation in regenerated plants. Genome 38:737–742

Zhang S, Warkentin D, Sun B, Zhong H, Sticklen M (1996) Variation in the inheritance of expression among subclones for unselected (*uidA*) and selected (*bar*) transgenes in maize (*Zea mays* L.). Theor Appl Genet 92:752–761

Zhang S, Cho M-J, Koprek T, Yun R, Bregitzer P, Lemaux PG (1999a) Genetic transformation of commercial cultivars of oat (*Avena sativa* L.) and barley (*Hordeum vulgare* L.) using in vitro shoot meristematic cultures derived from germinated seedlings. Plant Cell Rep 18:959–966

Zhang S, Zhang S, Cho M-J, Bregitzer P, Lemaux PG (1999b) Comparative analysis of genomic DNA methylation status and field performance of plants derived from embryogenic calli and shoot meristematic cultures. In: Altman A, Ziv M, Izhar S (eds) Plant biotechnology and in vitro biology in the 21st century. Kluwer, Dordrecht, pp 263–267

Ziauddin A, Kasha KJ (1990) Long-term callus cultures of diploid barley (*Hordeum vulgare*). II. Effect of auxins on chromosomal status of cultures and regeneration of plants. Euphytica 48:279–286

Subject Index

ABA (abscisic acid)   37
ABA treatment   50
aberrant callus cells   170
abscisic acid (ABA)   37
Ac-element   78, 80
Ac transposable elements   96
Agrobacterium   1, 3, 7–9, 23, 24, 26, 27, 31, 32, 40, 44, 45–49, 66, 67, 74–77, 84, 85, 105, 107, 144, 153, 154, 160, 161
agronomic equivalence   120, 121
agronomic performance   178
allergenicity, GM rice   143
allergenicity, pest-resistant corn   139
allergenic protein, inherent   148
alteration of expression   177
aneuploidy   169, 176, 177
Aqueoria   2, 11
Arabidopsis   2, 8, 12, 25, 26, 28, 32, 37, 39, 44, 59, 77, 83, 85, 87, 88, 159, 163
assessing GM crops   147
ATA method (aurintricarboxylic acid)   53
*att*P-flanked region   89–90, 95
autofluorescence, chlorophyll   32
autofluorescence, overlap with GFP fluorescence   23
auxin   115
Avena   170

Bacillus   122, 124
Bacillus thuringiensis (Bt)   148
bar gene, selectable marker   5
barley   169–183
beneficial insects   149
benefits *cry* proteins   123, 127
binary vectors   13
binary vector plasmid   114
bioluminescence   32
bioluminescent reaction, catalyzed by luciferase   31
biaphorus gene marker   141
biotechnology-derived foods, safety   119
birdÖs-eye view, scenario study   149
bombardment   174, 175
BP band-pass filter, cut off chlorophyll-red autofluorescence   26
BSA (bovine serum albumin)   50
Bt corn, performance   128–133

Bt toxin   148

Caenorhabditis   20
CaMV (cauliflower mosaic virus)   66–68
CaMV promotor   67, 68
cauliflower mosaic virus   66–68
cellular signal transduction pathways   39
certification program, genetic manipulation   65
chaconine   145
chimeric *ipt* gene   106
chimeric transgenic plants, generation   102
chloroplast DNA, diminishing gene flow   149
chromosomal aberrations, transgenic soybean   153–165
chromosomal variation   169–183
chromosomal fidelity   176
chronic toxicity examination   147
CIM (callus-inducing media)   171
cis-acting elements   37
Citrus   26
cloning vector   114
cointegrated vectors   14
commercialization transgenic crop   73
compositional equivalence, GM food   120, 121
concomitant GUS expression   87
considerations, ethical and religious   63
corn, pest-resistant   139
co-suppression   163, 181
co-transformation   14
co-transformation strategy   76
co-transformation system   95
co-transformation, with segregating T-DNAs   77
cotton, insect-protected, feeding studies   129, 133–134
cottonseed, feed containing *cry* protein   133–134
crops, genetically modified (GM)   139–149
Cre/loxP system   82–86
Cre-mediated excision   83
crown galls   7
cry insect-control proteins, safety assessement   124–125

cry (crystal) proteins 122–125, 127, 134
cry proteins, degradation 125, 126
cry proteins, mode of action 125, 126
cry-protein-specific receptors 125
CTAB-based method, DNA extraction 65, 66
cytokinin 7, 115
cytological instability, in transgenic plants 174

dehydration condition 49
dehydration gene expression 37–59
dehydration stress 38
dehydrative-responsive element (DRE) 37
desired genes, cloning vector 114
desynaptic gene mutations, sterility 162
detection equipment, GFP 22, 23
detection transgenic sequences 64
different GFP for specific uses 25
digestible-energy experiments, swine 133
digestibility, GM rice 141, 142
direct repeat (DR) structure 82
DNA, extractability 64
DNA, extraction 65, 66
DNA, fragmentation 64
DNA level, detection of genetic manipulation 64, 65
DNA microarrays 147
DNA probe, specific for target sequences 69
DNA transformation, polyethylene glycol mediated 161
dominant selectable markers 3–7
DRE (dehydration-responsive element) 37
DREB (DRE-binding proteins) 38, 57–59
DRE-independent process 39
Drosophila 20

electroporation-mediated transformation 161
electroporation protoplasts, soybean 154
elimination marker genes 73–91
endoreduplication 169
enhancer-trap construct 8
enhancer trap, promoter tagging 57
environmental safety 148
epigenetic gene mutation, sterility 162
equivalence, substantial 139
escape markers 1
Escherichia coli 2, 7, 39, 141, 142
excision marker, GFP 12
excision, transposon-mediated 88
excitation maxima, GFP 20
explants, genetic background 159
expression, stable 23–28
extraction control 68

farm-animal studies, insect-protected crops 128–133
fatty acids, in GM rice 143
fatty acids, potatoe 144
FDA (US Food and Drug Administration) 139
feeding value, testing 119–135
feed safety, testing 120–122
ferritin, soybean 147
Festuca 176
fidelity, transgenic plants 177–184
field performance 178–180
filter sets, GFP 26
first generation GM crops 140
FISH (fluorescence in situ hybridization) 164
floral dip, in *Agrobacterium* suspension 47
FLP-mediated recombination 87
FLP/*frt* system 86, 87
fluorescence *in situ* hybridization 183
fluorometric assay, GUS activity 49–51
food additives, safety assessment 143, 146
food industry 63
food safety standards 119–122
foods, biotechnologically derived 139
foreign proteins, expression 140
   GM crops 140
foreign proteins in single crop 146
fragment length 67, 68
fragment length, DNA 64
fumonisin 123, 124
functional proteins, dehydration 37, 38
Fusarium 123
fusion construct 13
fusion protein, expression 8

gate way cloning system 114
gene, artificially designed 144
gene expression luciferase 31–35
gene flow 148, 149
gene silencing, cytological basis 163, 164
genetic manipulation, detection 63–71
gene-transfer by infection 99, 104, 107, 109, 111
gene-trap constructs 8
genome stability 177
genomic alteration 176, 177
genomic DNA, from regenerated plants 178
genomic methylation 182
genomic stability 177–179
German Food Act 64
germination medium, *Arabidopsis* seeds 49
GFP (green fluorescent protein) 2, 11–13, 19–29, 31, 41
GFP-based selection 27, 28
GFP detection 24, 26

GFP expression 22–28
GFP, fluorescence intensity 28
GFP-positive cell mass 25
green fluorescent calli 23
β-glucuronidase 37–59
β-glucuronidase (GUD) assay 40
Glycine soja 139–149, 153–165
glycinin 141, 142, 144, 145
glycinin, expression 142, 144
glycinin, structure 142
glycoalkaloids, potato tuber 145
glyphosate-tolerant soybeans 156
glyphosate-tolerant transformants 153, 156
GM (genetically modified) 139
GM crops, gene flow 148
GM foods 140
GM material, relative percentage 70
GM potatoes 139
GM prime crops 146
GM rice grains 141
grain legumes 63–71
grazing preference, beef, insect-protected crop 132
GST-MAT vector 95–116
GUS (β-glucuronidase reporter gene) 9–11, 19, 21, 34, 39–46, 97, 98, 101, 106, 107, 112, 163
GUS assay 42
GUS construct 40
GUS expression 184
GUS expression, concomitant 87
GUS gene fusion 44
GUS reporter gene 58
GUS, histochemistry 51, 52
GUS, northern analysis 52–56
GUS progeny 76, 78
GUS, staining 10
GUS system, application 56, 57

heat-shock promotor 87
herbicide resistance, instability 163
high salinity condition 49
high-salt gene expression 37–59
„hit and run“ cassette 97, 113
homologous of DREB 58
homologoues recombinations 89, 90
Hordeum vulgare 169–183
hygromycin phosphotransferase 4
hypercholesterolemia 141

illegitime recombinations 89
imaging luciferase activity 32–34
immunological tests 147
insect-control proteins 122
insect-protected Bt corn 126, 127
insect-protected crops 119–135

insect-protected crops, growers acceptance 127
insect protection traits 122, 123
integration host factor (IHF) bacterial-encoded 89
intrachromosomal recombination 89
introduced proteins, safety 120
introns 32
inverted repeat (IR) structure 82
in vivo safety 145
in vivo safety, GM rice 143
ipt and iaaM/H genes, combination 108–110
ipt gene 81, 88
ipt gene, advantages 101
ipt gene, promotor 105–108
ipt-shooty 106
ipt-type MAT 112, 114
ipt-type MAT vectors 96–102, 103, 105
isopentenyl transferase (ipt) 80, 81
IVOMD (in vitro degestible dry matter) 132

kanamycin 3
kanamycin selection 40, 48
Klebsiella 2

LightCycler-techniques 70
loss-of-function analysis 45
luc (firefly luciferase) 11, 19–21, 31, 41
luc expression 32
luciferase activity, measuring 34, 35
luciferase expression 83
luciferase gen 31–35
luciferase standard curves 35
luciferin 32, 34
luminescence plateau, luciferin 34
luminometer 31, 32
luminometer, microtiter-plate 34
lysine content 140

marker-free plants, appearance 99, 104, 107, 110, 111
marker-free transgenic plants 74, 78–81, 84–86, 88, 90, 91, 102
marker-free transgenic plants, separation 102
marker-gene delivery 13–14
marker-gene elimination 85
marker-gene excision 82
marker genes from transgenic plants 95–116
marker genes, screenable 8–13
marker genes, selectable 73–91
marker, visually selectable GFP 20, 21
markers 1–14
markers, counterselectable 7, 8

markers, innocuous   7
markers, selectable   5–7
markers, useful selectable, transformation
    technology   21
MAT (multi-auto-transformation)   95–116
MAT cassette   96–100, 114
MAT vector   110, 115
Mendelian inheritance,
    in transformants   163
metabolic fluctuation   144, 148
metabolic fluctuation, GM rice   143
methylation polymorphism   178, 184
microprojectile bombardment   173, 174
MQ (Mary Queen), potatoe strain   144–146
multi-cloning sites   114
multiple stop codons   32
multiple T-DNA, co-transformation   74, 75
mycotoxins   123

NAM (naphthalene acetamide)   89
neomycin phosphotransferase   3, 4
Nicotiana   105
non-GM potatoes MQ   146
non-primary crops   146
non-target insects   148
nontransgenic callus   172
nontransgenic plants   170–172
nontransgenic segregants   179
nopaline   77
nopaline synthase geneterminator   144
northern hybridization, GUS analysis   55,
    56
novel foods   63
novel selectable markers   7, 8
nutritional wholesomeness, farm
    animals   122

octopine   77
OECD (organization for economic
    cooperation and development)   139
Oryza   113, 139–142, 170
osmotic treatment, cytological
    aberration   175
overexpression of DREB   59

particle-bombardement   3, 5, 12, 76, 86
particle-bombardment-mediated
    transformation, soybean   154, 159
PCR (polymerase chain reaction)   64
PCR analysis   65, 66
PCR, competitive   69, 70
PCR control   66–69
PCR inhibitors   66, 70
PCR primers, specific   64
PCR program   67
PCR, real-time   69, 70
PCR systems, specific   66–69

PEG (polyethylene glycol)   50
Pennisetum   170
pest-resistant GM potatoe   145
phage-attachment region   95
phenotypic abnormalities, transgenic
    plants   180
phenotypic variation   179
phosphinothricin acetyltransferase   4, 5
Photinus pyralis   2, 12, 31
pHPI 106 vector   96, 98, 99, 101, 104
physical transmission   180
plant transformation   73
plant transformation programs   91
plant transformation technology,
    avaluation   19
ploidy level, in transgenic plants   176
ploidy level, transgene expression   183
pollen movement   148
polyethylene glycol-mediated DNA
    transformation   161
Populus   105, 107
position effect   148
post-bambardment selection, transgenic
    lines   156
post-transformation stress   154
potato patatin gene-promotor   144
potatoes   139–149
potatoes, sugars   145
potatoes, GM   144–146
prediction function of genes   147
prediction study, birdös-eye view   149
primary transformed tissue, GFP
    detection   23–28
ProDH (rehydration-inducible proline
    dehydrogenase)   38
proliferation process, in vitro   177
promotor analysis, GUS system   43
promotor-GUS   42
promotor-GUS-constructs   41, 49
promotor-GUS fusion genes   44, 45
promotor-GUS-system   37, 38
promotor tagging, GUS system   57
protein assay, GUS extract   50
protein fluctuation, GM rice   143
protein property, GFP   20, 21
protoplast, electroporated   34
protoplast, transformation   3
protoxins   125

quantitative PCR systems   69

rapeseed   77
rearrangements, structural,
    chromosomes   169
recombinase   99, 116
recombinase-mediated excision   85
recombinations, site-specific   1

recombination strategies, cre-mediated
    site-specific 82
regenerability, *in vitro* culture 176
R-gene, promotor 103–105
regulation novel food directive 63
regulatory proteins, dehydration stress 38
R-mediated excision 88, 89
R-mediated genomic deletion 88
removal of marker genes 95–116
removable element 97
Renilla 31
reporter of gene expression,
    use of GUS gene 38
repositioning, transposon-mediated 78, 79
rice 77, 113, 139–149
rice actin promotor 184
rice transformation 1–14
risks, unintended 63
RNA blotting, nothern analysis GUS 54
RNA extraction, ATA method, GUS 53
RoundupReady soybeans 67, 69
R/RS system 88, 89, 97
RT-PCR (reverse transcription polymerase
    chain reaction) 41, 43

Saccharomyces 81
safety assessment 119–135, 147
safety assessment, GM 139–149
safety concerns 63
sample preparation, transgenic plant 50
screenable markers 1–14
screening for segregating populations,
    GFP 28, 29
SCV (somaclonal variation) 169, 170,
    177–179, 184
Secale 170
second generation GM crops 140
segregation analysis, transformation
    Arabidopsis 48
selectable markers 1–14, 73–91, 95
selectable markers, dominant 5–7
selectable marker genes, removal 102
selection treatment 175
selective agents 2
signal perception, stresses 39
signaling pathway, stress response 39
single step excision strategy 82
single-step transformation 112–115
site-specific recombination 90
site-specific recombination systems 81–89,
    95
small-subunit promotor, rubisco 13
solanine 145
somaclonal variation 169
somatic excision, GFP activity 12
soybean 139–149, 153–165
soybean lectin gene 66, 67

spatiotemporal expression profile,
    determination 8
spray, *Agrobacterium* suspension 47
storage proteins, potatoe 141
Streptalloteichus 2
Streptomyces 2, 3
stress condition 49
stress-inducible *cis*-acting element 45
stress-responsive gene expression 58
sub-acute toxicity 146, 147

TaqMan techniques 70
target sequence, 18S RNA, eucaryote-
    specific 66
T-DNA, gene targeting 181, 182
T-DNA insertions 78
T 1 generation, *Arabidopsis* 40
Ti plasmid 13
tissue-specific localization, GUS 40
tobacco 77
toxicity examination, chronic 147
toxicity, GM potatoes 146
toxicology testing, whole food 121
traceability, genetic manipulation 65
transactivation experiment,
    GUS system 57
transcriptional fusion marker 10
transcripts, transgenic plant 139
transformants, primary 161
transformants, stability 164
transformation 173
transformation, *Arabidopsis* 46–49
transformation, *Agrobacterium* 1, 3, 13,
    46–49, 77
transformation, electroporation-
    mediated 161
transformation, pretreatment with 2,4-D
    156, 158, 161, 162
transformation stresses 174
transformation technique 26
transformation tobacco leaf fragments 24
transformation, treatment 175
transformation vectors 78, 81
transformed seeds, T1 48
transformed shoots, detection 19–29
transgene expression 169, 181
transgene expression, loss 76
transgene expression segregation ratio 183
transgene expression stability 184
transgene inactivation 184
transgene integration sites 182
transgene silencing 181
transgene stacking 110–112
transgenes, co-delivery 13
transgenes, cointegration 14
transgenes, stability 180–184
transgenic crops 73–91

transgenic cultivar   91
transgenic plants, fertility   3
transgenic plants, field release   73
transgenic plants, preculture   113
transgenic plants, regeneration   99, 104,
    107, 109, 111, 172–177
transgenic shoots   107
transgenic soybeans, aberrations   153–165
transgenic soybean, chromosomal
    analysis   155, 160
transgenic soybeans, seed fertility   160–163
transgenic tissue, GFP marker   23
transient assay, GUS system   56, 57
transient expression, GFP detection   22
transient GFP detection   27
transient GFP expression   27
transposable element   96, 97
transposase   78
transposon-mediated relocation   78–80, 88
transposon-mediated strategies   90
Triticum   170

troubleshooting, GFP detection   22, 23
"tumor rooty"   7
two-step transformation   95, 100, 103–112
two-step transformation method   113, 115

ubiquitin promotor   12, 184
unexspected segregation   163

vacuum infiltration, *Agrobacterium*   47
variation, somaclonal   169
vector gene   144
vectors, GUS gene fusion   44
Vicia   139

wholesomeness, nutritional   120
working solution, luciferin   33

X-Gluc reaction   41

Zygosaccharomyces   81, 97

Printing (Computer to Film): Saladruck Berlin
Binding: Stürtz AG, Würzburg